NORTHWEST

Regional Environmental Issues Manual

Bringing Environmental Issues Closer to Home

Barry Thomas

California State University at Fullerton

SAUNDERS COLLEGE PUBLISHING

Harcourt Brace College Publishers

Fort Worth Philadelphia San Diego New York
Orlando Austin San Antonio Toronto
Montreal London Sydney Tokyo

Regional Environmental Issues Manuals are unique supplements designed to accompany Saunders College Publishing environmental textbooks including *Environment*, by Raven, Berg, Johnson, and *Environmental Science* by Arms. Each *Regional Environmental Issues Manual* is designed to promote grassroots awareness of local environmental issues, problem-solving analysis, and verbal and written discussion of topics that pertain to specific regions of the United States.

Regional Environmental Issues Manuals attempt to present a range of views on select environmental issues in a non-biased approach within limited space constraints. The goal of the manuals is to encourage student individual analysis of complex issues which go beyond the scope of the publication. Thought-provoking questions, commentary, and readings have been included to stimulate students to investigate the issue in further detail beyond the manual's presentation.

We invite your comments, ideas, and feedback. After reading the manual, please complete and return the Professor and Student Comments Form located in the back of the manual.

Acknowledgments
The project has been made possible with the help of my students; Susan Richards, Cheryl Solomon, George Wirtes, James Wilson, Michele Garden, Matt Atwood, and Amy Fields.
Credit must be given for their help in finding those interesting articles and preparation of the text.

Printed in the United States of America.

Thomas: Northwest Regional Environmental Issues Manual

ISBN 0-03-097142-X

345 256 98765432

NORTHWEST REGIONAL ENVIRONMENTAL ISSUES MANUAL

Table of Contents

Table of Contents

Northwest Profile

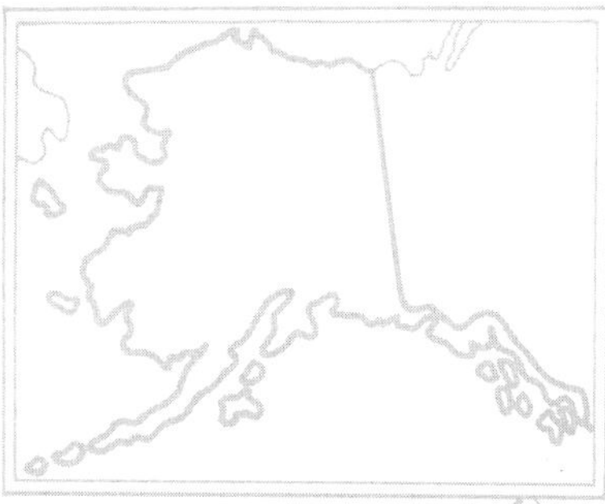

ALASKA
WASHINGTON
IDAHO
OREGON
MONTANA
WYOMING
COLORADO
NORTHERN CALIFORNIA

TABLE OF CONTENTS

1. Life in the Fast Lane of Our National Parks
by Susan Richards
This exercise investigates the National Park design controversy and the problem of getting rid of the developments and yet accommodate the crowds. Consider whether we should accommodate the tourist or do we maintain the primitive state of our National Parks? Two National Parks are highlighted,Yosemite National Park in California, and Denali National Park in Alaska. Our second concern is what happens to our wildlife in these situations? Wildlife concerns versus visitor demands. The problem: How can we limit the human-bear conflicts in Yellowstone National Park and how can we reintroduce the wolf to Yellowstone National Park in Montana?

2. Endangered in the Northwest
by Cheryl Solomon
This exercise examines the issues surrounding the controversy of the spotted owl, black footed ferrets and our amphibian populations of the Northwest. What is happening to them? Why are they having trouble and what can be done to remedy the situation?

3. Marine Mammals...Cute But With a Price
by George Wirtes and James Wilson
This exercise investigates the problems with two of our well known marine mammals, the sea otter and the sea lion. One is cuddly with a public image that is hard to beat and the other relies more upon the scientific data for its protection rationale. Both are fully protected but the conflict between them and people are putting strain upon the protected species laws.The sea otter is not just a small adorable animal, it is a feeding machine that consumes greater than 20% of its body weight in food each day. Favorite foods include bivalves, sea urchins, abalone, and crustaceans. The sea otter (*Enhydra lutris*), once had a range from the tip of Baja California to the Aleutian Islands in Alaska. Unfortunately, their pelt has been such a valued commodity that they were almost trapped into extinction by American and Russian fur traders. By 1914 the total otter population was estimated to be approximately 100 animals. Action was taken in 1911 in the form of an international treaty prohibiting the trade of otter pelts and in 1973 the effort was further supported by the U.S. Marine Protection Act. Today, the otter population in Alaska ranges from 100,000 to 200,000 animals and 1,600 animals in California. It is evident that the populations are recovering very well. They continue to slowly push their way south invading the territory they once occupied. With their expansion they bring several concerns to government, special interest groups, and private industry.

4. Logging and the New Deal
by Michele Garden
In this study the logging practices of the U.S. Forest Service comes under scrutiny. Sometimes the deals arranged by Government services are not in our best interest. In this case you be the judge. Should we stimulate the local economy at a high subsidy or should we protect the resource?

5. The Exxon Valdez Oil Spill
by Matt Atwood
On Good Friday, March 24, 1989, the fully loaded oil tanker, Exxon Valdez, ran aground on Bligh Reef spilling 232,000 barrels (11 million gallons) of crude oil into the pristine beauty of Prince William Sound, Alaska. The oil spread in a Southwesterly direction into the gulf of Alaska covering over 25,000 kilometers of coastal and offshore waters occupied by various species of plants and animals. The Exxon Valdez spill was not the largest spill in the world, but because of the dense population of animals inhabiting the sound, it quickly became one of the most devastating man made disasters in the history of the United States.

6. Hazardous Waste: Not in Our Backyard!
by Barry Thomas
This exercise examines the impending grid lock that is about to fall upon the researchers, medical laboratories, and energy producers of California. Your well being may depend upon the outcome of this environmental (or is it hysterical?) confrontation.

7. The Fight To Save Our Wetlands
Amy Fields
This is a study of what several of the western states are doing with their wetlands. Are they doing a conscientious job in protecting them and how are they dealing with such problems as maintenance, protection of seasonal wetlands, governmental cooperation, restoration of degraded areas and mitigation measures?

8. Gillnet Fishing on the West Coast
by Barry Thomas
This issue examines the attempt to protect our marine resources. The sports fishing industry has been vocal in trying to get attention to the fact that our fish are disappearing. Tuna cannot be found where they were once plentiful. The Gillnet initiative is a controversial means of solving this problem...or will it?

9. Rivers, Rivers, Rivers, and Not a Drop to Drink
by Barry Thomas
This exercise looks at the problems facing rivers of the Northwest. The conflicts between those who want to use this resource are a major problem pitting the economy against the environment and the need for power against the fisherman.

Introduction

Life in the Fast Lane of Our National Parks

This exercise investigates the National Park design controversy and the problem of getting rid of the developments and yet accommodate the crowds. Consider whether we should accommodate the tourist or do we maintain the primitive state of our National Parks? Two National Parks are highlighted, Yosemite National Park in California, and Denali National Park in Alaska. Our second concern is what happens to our wildlife in these situations? Wildlife concerns versus visitor demands. The problem: How can we limit the human-bear conflicts in Yellowstone National Park and how can we reintroduce the wolf to Yellowstone National Park in Montana?

Susan Richards

YOSEMITE NATIONAL PARK

Yosemite Valley is congested with more than a thousand buildings, stores, homes, garages, apartments, lodging facilities, and restaurants. The Valley floor is bisected by approximately 30 miles of roads which accommodate over 1 million cars, trucks, and buses a year. The intent of the National Park Service is to remove all automobiles from the Valley and to redirect development to the periphery of the park and beyond. There is a serious traffic problem, as many as 60,000 people a day at peak (6-7,000 cars a day), and the pollution is above all state and federal air quality standards.

The General Management Plan was devised from a workbook developed by the Park Service which was distributed to over 60,000 groups and the result was a mandate from the U.S. Congress. First, to protect and preserve the resource. Second, to make the resource available for public amusement and education. Thirdly, though not directly spelled out, to know when to draw the line between the two.

The Plan was designed to cut down congestion by removal of valley campsites and increasing park-wide sites; removal of park headquarters to periphery; to relocate employee housing; removal of over 1,000 parking spaces; lowering of day-use capacity; a shuttle bus service from periphery to interior; and eventual removal of all cars and obtrusive vehicles. Congress enacted a measure to acquire land in El Portal to help with the Plan, although little was done with it. The obstacles stated by the Park Services for using El Portal were not a serious constraint on development of El Portal according to Yosemite's chief scientist, Jan van Wagtendonk. Not only that, but Curry Co. has built or expanded more than a dozen "profit centers" since the enactment of the Plan. Moving NPS buildings from Yosemite Valley will partially fulfill General Management Plan (GMP). It is estimated to be complete in 1996 based on the availability of funds. But it still leaves one of the biggest problems, automobiles in the park. Inappropriate development has plagued Yosemite Valley for most of this century. Heavy auto traffic, motels, gift shops, liquor shops, and a video outlet just to name a few. Also, visitor numbers have increased by 37% in the last ten years. Some of the improvements made by the Curry Co. include a recycling program, removal of some houses and a golf course in Yosemite Valley, and support of the return of the peregrine falcon and big-horned sheep to the backcountry.

Last year, Manuel Lujan, Secretary of Interior, went public with his doubts about the appropriateness of foreign ownerships of National Parks concessions after Matsushita Electric Industrial Co. bought Curry's parent co, MCA. Lujan has accelerated concession reforms. In January, the firm agreed to sell Curry Co. to NPFund for $49.5 million. Lujan also wants franchise fees to be raised to 22% of gross receipts with 5% going back to general treasury, and 17% going back to the park. Instead of the 3/4 of 1% gross receipts they pay now.

DENALI

After seven years of debate, in 1980 protection of 100 million plus acres in Alaska was finally settled. This doubled the size of the National Park System. For years Alaska was being torn apart between the claims of those who want to protect the natural splendor and those who are pursuing the possibility of profit. The park wants to prevent overcrowding, excessive development and air pollution. The Management plans are outlined as follows:

1. Limiting the development of lodging and other visitor services to areas outside the park.
2. Open Alaskan park land to trophy hunters and trappers (bill was defeated).
3. Preexisting right of ways to the park roadways, rail lines, and certain undeveloped areas will remain intact.
4. All mining operations in Alaskan parks must cease on October 15, 1985 until the mine operator submits its plans to the NPS, which will produce environmental assessments for each plan and enforce these regulations;

In 1990 a private landowner, Dan Ashbrook, proposed a plan to build an RV park within Denali National Park. This would pose long term threat to the wilderness appearance. This would also mean that the visitors would have to drive through the park (right now they have a bus and shuttle system). This would also seriously disrupt the wildlife habitat and sightings. The way the park is now it has 5 private lodges with small scale development and the guests all travel by bus. The addition of the RV park would be very detrimental to the beauty of the park but would be welcomed by many tourists, particularly old folk and mobility handicapped.

In 1992, a task force agreed that the current access into the park combines wilderness protection and visitor access better than the proposed alternative. The task force suggested improvements but was against any major new road construction. Denali is the most visited park in Alaska due to Mount McKinley (which is the highest mountain in North America) and its unparalleled wildlife (high abundance and ready visibility). The winding 90 mile gravel road is on a system of shuttle buses that provide safety and excellent wildlife viewing, by limiting traffic and disturbing the animals as little as possible. Last year it was proposed to add a second road (paved concrete) to expand tourism, but the task force advised against the proposal.

YELLOWSTONE

Human-Grizzly bear conflicts are a problem at Yellowstone National Park and in March 1993 a grizzly attacked and ate a tourist in Jasper National Park, Canada. One of the major reasons for this problem is the Grizzly's loss of habitat due to development and tourist pressures. The Grizzly now inhabits less than two percent of its original range in the lower 48 states. The numbers have decreased from 50,000 to 1,000 animals. The U.S. Fish and Wildlife Service says that not enough is being done to protect bears and this will push them closer to extinction. It is feared that as their habitat decreases they will split into smaller isolated populations without the genetic diversity needed to survive.

Decreasing their habitat also results in decreasing their foraging area. This leads to the bears trying to obtain food from visitor campsites and poses a serious threat to visitor safety. The Interagency Guidelines for Management of Grizzlies in the Greater Yellowstone Area sets forth specific criteria on what constitutes a "nuisance bear" and on how to deal with them if the bear is seen repeatedly near developed park areas:

1. Aversive conditioning – scare them away with rubber bullets, projectiles and noisemakers.
2. Next, trap and relocate the bear to back-country.
3. Finally, remove the bear from the park permanently by placing it in a zoo or shooting the bear.

One of the many incidents Yellowstone has recorded the most since 1967, occurred on October 4, 1986 when William John Tesinsky (aged 38) was killed by a bear sow (aged 8) labeled Griz number 59. This bear had been exiled from the park twice and returned both times (once from over 25 miles away). Griz number 59 was destroyed upon being found with a human carcass. The facts and circumstances indicated that Tesinsky had approached the bear too closely and possibly provoked the bear, and therefore, was attacked and killed. The reason given for killing the otherwise non-aggressive bear was the theory that once a grizzly has attacked and killed a person, or evenly seriously injured one, once it has tasted manmeat – it is likely to attack again. Due to this prevailing theory in the lower 48 parks, the errant bear is more often destroyed than it is spared.

There have been several different proposed ways to handle this problem without having to kill the bears. The Interagency Guidelines for Management of Grizzlies in the Greater Yellowstone Area is just one. Yellowstone officials advise visitors to use extreme caution when they observe or encounter wildlife. They also advise to wear loud bells on their shoes when hiking, and to secure their food out of the reach of the bear. The park crews also clear away the underbrush around trails or put up electric fences (not very effective) to prevent attacks. Their main concern is for the safety of park visitors and the protection of Yellowstone's wildlife and the habitat which supports it. Some biologists believe that we can recondition the trouble causing bears so that they can remain in their natural ecosystems and no longer threaten people and property. Every Grizzly removed from the Yellowstone ecosystem represents a serious loss to a threatened species (protected by the Endangered Species Act since 1973). And removal of a reproducing female is even worse.

Wolves are yet another concern. Many biologists and concerned citizens wish to re-introduce the wolf to its original range. Supporters of restoring wolves to Yellowstone greatly out numbered opponents at the U.S. Fish and Wildlife hearings held in August 1992. Some 1500 people attended the meetings held in six western cities. Some objections to this new round of hearings were voiced by the U.S. and Wildlife Service because of the cost and the fact that these hearings tend to polarize the issue. Proposed re-introduction should take place in 1994. It is estimated that increased tourism in Idaho, Montana and Wyoming would be in the $43 million range . . . all because of a few wolves! Now that is a financial impact that interests many people.

Reading 1

Questions and Concerns

READING 1

When reading this article consider the following:

1. Where is this park?

2. Who goes there and why?

3. Consider the old folk and handicapped persons, should they not be able to see the sights in some comfort? Is a yellow school bus adequate transportation?

The Taming of Denali

With tourism growing, Alaska's premier national park faces development pressures.

Story by Kim Heacox

AMONG THE RIDDLES of the ancient Chinese is one about a man who discovers the most beautiful place in the world. The riddle is a conundrum—a puzzle without a solution—because the man shares his discovery, and people flock there in such great numbers that the place is changed forever and is beautiful no more.

It is the pioneer's paradox, the process of people destroying, or at least eroding, the very thing they love—usually the natural environment—and it exists not just in ancient China but throughout the entire modern world.

National parks should be exempt from this paradox, but unfortunately they are not.

This is the story of a great national park teetering on the edge of that dangerous precipice. Will the same mistakes made elsewhere be made here? Or will people learn that to truly save a place they must close doors in front of them rather than behind them?

The park is Denali—the most popular, visible, and accessible national park in Alaska. Decisions made here may likely set the course for parks in the rest of the state for a long time to come. One hopes that Alaska can avoid the Manifest Destiny mentality that fenced and tamed the lower 48 states; that we who can alter any landscape in the world will have the wisdom to leave this one alone. So far, the prognosis is uncertain.

It is vital that people come here; that lives are touched and inspired, that wilderness values are affirmed and anxieties washed away. Yet it is equally important that visitors not be herded into overcrowded parks because of arrogance and greed.

At 6 million acres, Denali National Park and Preserve is about the same size as Massachusetts, nearly three times as large as Yellowstone. Running through it are icy mountains that break their backs in the Alaska Range, their summits reaching to 14,000 feet, 17,000 feet, and finally to 20,320-foot Mt. McKinley, more properly called Denali—the Native name meaning The High One—the highest mountain in North America. From the mountains, the land sweeps to every horizon in striking patterns of tundra and spruce forest, kettle ponds and braided rivers, wildflowers and willow thickets. More than 450 species of trees, shrubs, and herbs live here, some growing profusely in protected valleys, others hugging the earth in button, mat, and rosette shapes atop windswept ridges.

From around western North America and the Pacific Rim—Siberia, Japan, Hawaii, California, Central America, South America, and Antarctica—birds come here to raise their young. Shorebirds nest on the tundra, raptors on the cliffs: more than 150 species occur here. But the most visible and sought-after residents are the mammals: the grizzlies, wolves, caribou, moose, Dall sheep, red foxes, beavers, arctic ground squirrels, lynx, showshoe hares, pikas, and many others: 37 species in all. Nothing stimulates the heart more than the sudden appearance of a bear, a wolf, or a herd of caribou moving like poetry across open ground. Like the birds and plants, they fit into the landscape as an integral part of a greater whole, manifesting laws of survival and diversity, helping to create what has been called "the greatest subarctic sanctuary in the world."

This, then, is what Charles Sheldon found when he came to Interior Alaska in the summer of 1906. A member of the influential Boone & Crockett Club, he was cut from the same conservation cloth as Teddy Roosevelt. A hunter, yes, but also a competent and caring naturalist who traveled widely throughout Denali by foot, showshoe, and sled dog team.

Camped on a moraine above the Peters Glacier in January 1908, with the land and the silence all to himself, he wrote: "When Denali Park shall be made easy of access with accommodations and facilities for travel...it is not difficult to anticipate the enjoyment and inspiration visitors will receive."

Eighty-four years later, Sheldon's prediction has come true. Enjoyment and inspiration are commonplace among visitors to Denali. But for those who feel the nascence of exponential growth; who hear the whines of an insatiable tourism industry always hungry for another hotel; who deplore the pro-development Alaska congressional delegation and respect the fragility of the subarctic web of life, Sheldon's crystal ball looks more like Pandora's box.

In 1971, the year before the highway was completed between Anchorage and Fairbanks, connecting Denali to Alaska's two largest cities, annual park visitation was 30,000. Today it is 600,000. A single dirt road, built in the 1920s and '30s, winds 90 miles through the park, cutting into mountainsides, crossing rivers, traversing open expanses of tundra and spruce forest. An average of 35 shuttle buses and 25 tour buses rumble over that road every day of summer, each carrying about 40 people who admire the scenery and watch for wildlife.

The bus system works on two premises: by reducing private vehicle traffic along the road, it minimizes the risk of accidents and maximizes the opportunities to view wildlife that otherwise might be displaced by more traffic.

The road ends at Kantishna, a mining district in the heart of Denali that has been a burr under the saddle of more than one park superintendent. Back in 1903, three years before Charles Sheldon arrived, gold was discovered her by a mountaineering party led by Territorial Judge James Wickersham. Within two years the Kantishna Hills were swarming with prospectors.

When Mt. McKinley National Park was created in 1917 (largely through the efforts of Charles Sheldon, who feared prospectors would overhunt Dall sheep and other wildlife), Kantishna bordered its northern flank. In 1980, with passage of the Alaska National Interest Lands Conservation Act (ANILCA), 2-million-acre Mt. McKinley National Park became 6-million-acre Denali National Park and Preserve. And Kantishna became an enclave.

At the time, John Cook, regional director of the National Park Service in Alaska, said, "There are two things you never want to see being made: sausages and boundaries." His comment was aimed at the political process more than the product, and applied to every new park and preserve created by ANILCA. Certainly ANILCA was a great accomplishment—the Louisiana Purchase of the American conservation movement—for it more than doubled the size of the U.S. National Park System. But there were compromises. Buried in the convoluted vernacular were the terms "reasonable access" and "traditional use," put there by the Alaska congressional delegation to protect the rights of people they called "honest, hard-working Alaskans who have nothing to gain with a land lock-up." Among those Alaskans: the miners of Kantishna.

In Denali, reasonable access and traditional use meant Kantishnans had the right to drive the park road. As the mines became less profitable and tourism more so, a few Kantishnans opened roadhouses, restaurants, and lodges. And though ANILCA did not require it, they nevertheless agreed with the National Park Service to minimize traffic and shuttle their clients in buses. A relative peace reigned until the spring of 1990, when a Kantishnan plowed off a gravel clearing, called it a motorhome park, and invited the public at large.

From a legal standpoint, it was permissible. But from a safety standpoint, mixing motorhomes with buses on a narrow, winding road was insane. Superintendent Russ Berry, having arrived in the park less than a year before, countered with a chess move. He reasoned that if one Kantishnan could have his clients driving their private vehicles on the park road, then so could the others. All the clients needed was a reason to go the Kantishna: a night's lodging, a meal, a cup of coffee. But to mitigate the dangers from increased traffic—and this is where Berry played his king—the National Park Service threatened to turn around every tour bus and shuttle bus at Teklanika Campground, only one-third the way out the road, long before the best views of wildlife and Mt. McKinley. Suddenly Kantishna had the entire Alaska tourism industry on its back.

When the wit and wind finally died down, the motorhome park was forestalled and schedules returned to normal. Yet pockets of Kantishna remained defiant, as they always have, peppered with the likes of Tom Anthony, who threatened to shoot anyone who crossed his property on the road, and the Wheeler brothers, arch-anarchists and Alaska's self-described "foremost authorities on recreational bulldozing," who more than once have threatened to cut a giant "W" on a mountainside in the park.

And what of the fruits of their labors, and of the nearly nine decades of miners in general in the Kantishna area? They pulled out a lot of gold. They also fouled a dozen drainages. Heavy metals—iron, arsenic, and lead—were oxidized and released at accelerated rates. Mercury, used as an amalgam, was dumped indiscriminantly. Barrels of petroleum products were left lying around. High-pressure hoses, the tools of the modern hydraulic miner, blasted hillsides. Erosion and siltation destroyed aquatic primary productivity. Streams balanced over the millennia, poetic in their finely balanced gradients and ratios of pools to riffles, were raped.

Some of the most serious damage happened in the early 1980s when gold prices rocketed. Since then, a few claims have voided, most have gone idle but are still valid. To purchase these lands, Russ Berry estimates he needs $60 million. So far, he's received one-tenth of that amount. "If the money arrives a little at a time, which it probably will," he says, "then ten years from now the remaining lands in Kantishna will probably still cost $60 million."

Meanwhile, the pressure grows to increase tourism and improve access into Kantishna. On July 13, 1991, U.S. Senator Frank H. Murkowski wrote a sophomoric newspaper editorial that began, "Question: Why does the Anchorage Zoo get more visitors each year than the entire interior portion of Denali National Park and Preserve, Alaska's premier visitor destination?

"(a) The popcorn is better at the zoo. (b) Denali doesn't have an elephant. (c) The zoo doesn't have an armed ranger to keep visitors out.

"The correct answer is the old standby, 'none of the above,' but answer (c) comes too close for comfort.

"It is becoming fashionable to say Denali National Park is being damaged by overvisitation and is losing its luster as a result. But is Denali National Park a victim of its own success, or has something gone awry in how it is managed?" Murkowski pointed out that Yellowstone receives a vastly greater number of visitors each year and is doing fine, then ended his sophistry by suggesting that an "elevated-rail transport system" be built "along the old Stampede Trail that runs east from Kantishna to the Healy area."

Here, then, is a way of thinking that believes Denali National Park should be accessible to as many people as possible; that tourism, like cattle ranching, is a volume-driven meat market; that scenery is, more than anything else, a commodity.

"It is not the job of Denali National Park to be the No. 1 visitor destination in our state," countered Mary Grisco, NPCA's Alaska regional director. "[We need] to educate people about what our national parks in Alaska offer and to let people know that [all] national park units do not provide the same experiences and amenities."

If the finest hotels can have "no vacancy," if the greatest concert halls can have limited seating, then why not our national parks? "The theater is full," says Russ Berry. "You are invited to the next performance." Or, as Aldo Leopold wrote in *A Sand County Almanac,* "It is the expansion of transport without a corresponding growth of perception that threatens us with qualitative bankruptcy of the recreational process. Recreational development is a job not of building roads into lovely country, but of building receptivity into the still unlovely human mind."

Seventy years ago, the only thing that ran through the Nenana River Canyon, next to the entrance of the park, was the Nenana River. Then came the Alaska Railroad. Then a major highway. Then a lodge, a motel, a cluster of chalets, restaurants, pizza parlors, a salmon bake, a beauty shop, river rafting offices, and helicopter pads. Last summer, four helicopters arrived, each carrying tourists into the park 12 times a day.

Soon the quality of the air and water goes down, wildlife movements are impeded, all-terrain vehicles rip in and out of the park, and moose are shot only 200 feet outside the boundary. It happened this fall. Denali is beginning to look like parks everywhere else: beleaguered.

Addicted to growth, tour companies add rooms and beds to chalets and hotels, then whine for more buses to accommodate their expanding clientele. Bus seats are oversold every summer, backcountry units fill up, and long lines form in Denali's visitor access center. "The truth is," says one ranger, "that in every major visitor survey the public has strongly supported the existing policies and level of development. More is not better."

Yet the incrementalism marches on. The tour operators and concessionaires make more money—their prime objective—while the visitors themselves are herded into mediocrity, onto tour buses at 5:30 in the morning that whisk them into the park and back out in time to catch the afternoon train north and make room for the next wave.

It is not another Yellowstone, but Denali is not what Charles Sheldon found in 1906, either. A new hotel is planned inside the park to replace the existing one at the same capacity, but at a price (approximately $25 million, paid by the federal government) and style that has stirred up stiff opposition. Does Denali need a grand, opulent hotel? Wouldn't the money be better spent elsewhere, such as purchasing inholdings in Kantishna?

South of the park, Native corporation land near Talkeetna is under consideration for a major hotel and National Park Service visitor center. Again, there is opposition, this time from Talkeetna residents who like their peace and quiet. And yet another visitor center is planned for High Lake, in Denali State Park, just south of the national park.

There have been victories: the establishment of a bus system to limit traffic and improve viewing on the park road, the creation of management units to control backcountry impact and protect sensitive wildlife zones, the beginnings of land acquisitions in Kantishna, and an overall enlightenment amid a growing environmental movement. But is it enough?

In a sweeping, grandiose state where the words "Last Frontier" carry the old, false assumptions of limitless resources and opportunities; where a congressional delegation and state government embrace economic growth as though it were a religion; where the National Park Service must answer to these same politicians; and where industrial tourism advances slowly and inexorably, like the tide; if this is the way it is, and shall be, then Denali is doomed.

Something has to change. A growing audience advocates removal of the National Park Service from the Department of the Interior and the creation of an autonomous governing body with a director who answers to a rotating board of distinguished American scientists, teachers, managers, writers, and artists. Lines must be drawn and defended, for only then will landscapes beat to the rhythms of something more ancient than us all. Here, they will say to themselves, is a piece of the earth as it once was and should forever remain: absolutely wild.

Kim Heacox writes and photographs frequently for national publications from his home in Alaska. He has authored several books and is completing two more, Iditarod Spirit *and* In Denali.

To express your concern about development of Denali, please write to Alaska's congressional delegation: Senators Frank Murkowski and Ted Stevens (U.S. Senate, Washington, DC 20510) and Rep. Don Young (U.S. House of Representatives, Washington, DC 20515). Send copies of your letters to Mary Grisco, NPCA's Alaska director (P.O. Box 202045, Anchorage, AK 99520).

Questions and Concerns

READINGS 2 & 3

When reading these articles consider the following:

What is a wolf, is it a killing machine or an animal bent on surviving?

Activists Cry "Pro-Wolf"

Endangered Species Act supporters flex muscle at public hearings held to consider the reintroduction of the gray wolf to Yellowstone

Like countless other endangered species, the gray wolf's plight reminds conservationists of the uphill battle we face to save endangered animals. Almost two decades of inaction have passed since the wolf was placed on the Endangered Species List in 1973. In the interim, the gray wolf's stealthy lope has nearly disappeared from the landscape of the lower 48 states.

While there has been modest success in Minnesota, efforts to restore gray wolf populations elsewhere have languished in red tape. The U.S. Fish and Wildlife Service took 14 years simply to finalize a recovery plan only to have the politically powerful western livestock lobby prevent the plan's implementation.

Now, fortunately, a large group of conservationists have formed a powerful contingent favoring the prompt reintroduction of gray wolves. The territory stalked for this battle is Yellowstone National Park.

Federation members and other conservationists appeared in full force at six August public hearings held in Wyoming, Montana, Washington, Utah, Idaho, and Washington, D.C. The hearings were held prior to the drafting of an environmental impact statement considering wolf reintroduction in Yellowstone. Of the more than 400 individual who testified for the record, more than 75 percent proclaimed themselves pro-wolf.

"The turnout proves the tremendous support that exists not only for wolves but for the Endangered Species Act as well," said Tom Dougherty, the Federation's Western Region Staff Director. "It is heartening to see the number of people who showed up to combat the so-called wise-use movement."

Tim Stevens, an Endangered Species Western Region Organizer for the Federation, attended hearings and rallies in Cheyenne, Wyoming, and Salt Lake City, Utah. Contrary to his expectations, the opposition produced only a paltry showing.

"This was an incredible victory for conser-

Reprinted by permission from EnviroAction, published by the National Wildlife Federation, 1400 16th St. NW, Washington, D.C. 20036-2266

vationists," Stevens reported. "The hearing record was overwhelmingly ten to one pro-wolf."

The western livestock industry and other anti-environment interests lead opposition forces. They are pushing for a "no Wolf Alternative" that would remove wolves from the Endangered Species List. This would allow the unregulated killing of wolves by the public and silence the animal's sonorous howl.

Wildlife biologists and other experts believe gray wolves can serve a vital function in the greater Yellowstone ecosystem. By culling sick and weak animals, these predators keep elk, deer and moose herds healthy and strong. Without wolves, prey populations unnaturally balloon—unbalancing the ecosystem and taxing wildlife management resources.

After a 60-year absence of the gray wolf, Yellowstone can only benefit from its return. As conservationists know:

- Gray wolves are not rogue killing machines. They prefer natural prey and rarely attack husbanded livestock. In Minnesota where 1,500 wolves live in proximity to 325,000 cows and sheep, only 170 (0.05 percent) are claimed to be wolf-killed annually.

- A private fund exists to reimburse ranchers for all verified losses to wolves.

- Wolves maintain a balanced ecosystem.

- Wolves are shy and avoid contact with humans whenever possible. There has never been a documented death caused by a wild non-rabid wolf in North America. In contrast 11 North Americans were killed by domestic dogs between May 1985 and September 1986.

"Those fighting against wolf reintroduction are working on a much broader agenda. They are just anti-Endangered Species Act," Stevens said. "This is a political issue. The multiple abuse groups are trying to use the wolves as scapegoats."

While conservationists have won this round, the influence that opposition groups wield should not be underestimated. Continuous pressure is essential to save endangered species languishing on the brink of extinction.

— *Jeff Burdick*

HOW YOU CAN HELP

You can help by writing to be placed on a federal mailing list to receive information about the Yellowstone National Park and the up-coming central Idaho Gray Wolf Environmental Impact Statement (EIS). The draft EIS is due by May 13, 1993, by the U.S. Fish and Wildlife Service.

Yellowstone National Park and
Central Idaho Gray Wolf EIS
P.O. Box 8017
Helena, MT 59601

Reading 3

WILDLIFE

Wolf Reintroduction on Track

Fish and Wildlife Service takes first step

The U.S. Fish and Wildlife Service (FWS) has finally taken the first important step toward reintroducing the endangered grey wolf into Yellowstone National Park.

Because the grey wolf is listed as endangered in forty-seven states and threatened in Minnesota, the Endangered Species Act (ESA) requires the FWS to work toward the wolf's recovery. In the 1987 Northern Rocky Mountain Wolf Recovery Plan, the FWS identified Yellowstone National Park as one of three key wolf-recovery areas and outlined the steps required to bring wolves back to Yellowstone. Little was done, however, until Congress acted last year.

In October 1991 Congress provided the FWS with $498,000 for the preparation of an Environmental Impact Statement (EIS) that will identify and analyze alternative plans for wolf recovery at Yellowstone and in central Idaho, another area earmarked for wolf recovery.

Before the arrival of European settlers, the grey wolf ranged over most of North America. Persecuted by settlers and ranchers, slowly starved as its prey—elk, deer, and bison—disappeared, the wolf all but vanished from the United States by the early twentieth century. Today's wolf population in the lower forty-eight states consists of only about 1,500 wolves in Minnesota and a total of perhaps five or six dozen individuals spread across Montana, Idaho, and other northwestern states.

The HSUS strongly supports the reintroduction of wolves into Yellowstone. We believe it is critical to both wolf conservation and restoration of a complete, healthy Yellowstone ecosystem. We immediately joined the EIS process, participating in April 1992 meetings intended to help identify the range of issues to be addressed in the EIS.

But reintroduction will not suffice to guarantee recovery. The ESA strictly prohibits killing, harassing, or otherwise harming members of endangered species. The FWS, however, will be pressured to relax the restrictions, for example, by designating wolves a "nonessential experimental population." Such a designation (permitted, under some circumstances, by the ESA) would allow the FWS to write special regulations that could weaken ESA protection for individual wolves.

For the sake of the recovery effort and the wolves' own safety, The HSUS believes that wolves reintroduced into Yellowstone must receive the full protection of their endangered status under the ESA. This need is underscored by the recent tragic history of the small wolf population occupying Glacier National Park and nearby regions of Montana. In the last two years, at least nine wolves have been killed—either deliberately and illegally or accidentally (in automobile collisions or other human-related accidents). To thrive, Yellowstone's wolves will need the strongest protection the law can give.

Wolves belong in Yellowstone National Park. The HSUS will do everything possible to help them get there and stay there. —*Allen T. Rutberg, Ph.D., HSUS senior scientist, Wildlife and Habitat Protection.*

Discussion for Life in the Fast Lane of Our National Parks

1. If you were a park ranger at Yellowstone how would you protect the bears and the visitors against possible threats?

2. Do you think that the rehabilitation of the bears will work? Why or why not?

3. Do you think it is fair that the bears lose their lives after a provoked attack?

4. Do you agree that a new paved road should not be built at Denali?

5. Devise a way to satisfy Dan Ashbrook (RV Park proposer) and the park officials and natural habitat. Can it be done?

6. Who would support the claim that the only good wolf is a dead one?

7. Should all National Parks have the original complement of animals? Or can we have specialized Parks?

8. Draw up a plan for specialized parks where different animals and dangers are presented.

   In which of our current National Parks would you put Wolves, Grizzly Bear and Elk?

Introduction

Endangered in the Northwest

This exercise examines the issues surrounding the controversy of the spotted owl, black footed ferrets and our amphibian populations of the Northwest. What is happening to them? Why are they having trouble and what can be done to remedy the situation?

The Pacific Northwest is a diverse ecosystem of mountains, deserts, marine shores and forests. The latter providing a major revenue source and an employer of thousands. Recently, because of the spotted owl controversy, the old growth forests have received a lot of media attention. Conservationists are trying to preserve much of the remaining habitat and the logging industry is bent on maximizing their investment. Workers and investors are adamant in their opposition to the preservation of old growth forests to protect an owl.

The spotted owl is not the only species being impacted by the clear cutting of the forests. At least five threatened or endangered animals cohabit the spotted owl range. These are the gray wolf, grizzly bear, Columbian white-tailed deer, bald eagle and peregrine falcon. The lynx, wolverine, mountain beaver, white footed vole, Pacific western big eared bat, marbled murrelet, bull trout, inland red-band trout, Olympic mudminnow, northwestern pond turtle, tailed and California red legged frogs and Del Norte, Larch Mountain and Siskiyou salamanders are other species whose numbers are diminishing due to the massive deforestation.

In 1990 it was estimated that 90% of the Northwest's old growth forests had been consumed by the timber industry. Smaller habitats mean smaller gene pools and an increase in survival risks. The spotted owl is seen as an indicator species. That means if the forests are managed to ensure the survival of that species then the other species will benefit. Is this sound reasoning? Can we make such a generalized statement or is this just a practical way of attacking the problem? Other animals that can be considered indicator species are the pileated woodpecker and the marten. Like the owl the pileated woodpecker nests in big trees and forages for insects in rotting logs. However, the pileated woodpecker can be found in the urban area and is no stranger to the suburbs of Vancouver. It is perhaps a little more resilient than is the owl. The marten dens in cavities of big trees and generally avoids clear cut areas. It requires a larger home range than the owl and therefore may be even more sensitive to clear cutting. But it is hard to see and does not have the same public relation image as does the owl.

Amphibians are disappearing in this region. Biologists have documented a dramatic decline in the frog population of the Cascade Range. In one study done at Lost Lake it was noted that nearly all frog eggs never hatched. While no distinct cause has been identified it is obviously a cause for alarm. Frogs may be a good indicator of environmental quality since they live both on land and in water. Their permeable skin makes them vunerable to many pollutants and pathogens in the environment. Some possible theories of the poor egg success mention acid rains, ultraviolet radiation, and global warming.

Although the Northwest has its share of species survival problems it does have one success story, the Black Footed Ferret. Once only 18 individuals survived, today, thanks to a captive breeding program in Wyoming, more than 300 are alive and well. Reintroduction has begun!

The following articles provide a more in depth look at these problems. Legislation concerning these animals is fairly recent and changes often. The Fish and Wildlife Services in each state can provide current lists, and the status of the endangered and threatened species in their jurisdiction.

Cheryl Solomon

Questions and Concerns

READING 1

When you read this article consider the following:

This paper was written to exemplify the role disease can play in efforts to preserve endangered species. Note the various preventative techniques used in this project. What group or groups of people do you feel the author wanted to influence in writing this article?

Reading 1

Disease and Endangered Species: The Black-footed Ferret as a Recent Example

E. TOM THORNE*

Wyoming Game and Fish Department
Research Laboratory
Box 3312 University Station
Laramie, Wyoming, U.S.A. 82071

ELIZABETH S. WILLIAMS

Department of Veterinary Science
University of Wyoming
Laramie, Wyoming, U.S.A. 82070

Abstract: *Diseases may play major roles in the conservation of endangered species. Although the threat of disease received extensive consideration and influenced research and management activities governing the endangered black-footed ferret* (Mustela nigripes) *in Wyoming, a canine distemper epizootic in 1985 severely affected a captive breeding program and led to extirpation of the species from the wild. This recent example of the catastrophic effect of epizootic disease in an endangered species is described in an historical context. In addition, examples are given of disease further endangering other rare species, including Mauritius pink pigeon, Père David's deer, cranes, maned wolves, native Hawaiian birds, cheetahs, and others.*

Resumen: *Las enfermedades pueden jugar un papel importante en la conservación de especies en peligro de extinción. Aunque la amenaza de enfermedad recibió gran consideración e influyó en la investigación y las actividades de manejo del hurón de patas-negras* (Mustela nigripes), *especie en peligro de extinción en Wyoming, en 1985 el distemper canino epizoótico impactó severamente un programa de crianza en cautiverio y condujo a la extinción de esta especie en su ambiente silvestre. Este reciente ejemplo del efecto catastrófico de una enfermedad epizoótica sobre una especie en peligro de extinción es descrito en un contexto histórico. Además, se dan ejemplos de enfermedades que amenazan aun más a otras especies raras, incluyendo la paloma rosada de Mauricio, el ciervo de David, el lobo de crin, el Cheeta, diversas grullas, varias aves de las islas de Hawaii, entre otros.*

Introduction

Although the literature presents numerous reports of disease-related mortality among free-ranging and captive wild animals, the occurrence of disease and its effects on threatened and endangered species is only infrequently documented. Yet, it is among these vulnerable species that disease may have major consequences for conservation of the species, including extinction. There are, however, a few documented epizootics and epornitics among threatened and endangered species. Many of these have involved captive animals, compromising captive propagation programs through decreased numbers of breeders and loss of genetically valuable individuals. The intent of this

Correspondence and requests for reprints should be addressed to this author.

Courtesy of CONSERVATION BIOLOGY, Vol. 2, No. 1, March 1988

paper is to describe the events leading up to and surrounding a case in which disease played a major role in the management and conservation of an endangered species and to provide examples of other diseases that may affect endangered species conservation programs.

Canine distemper is a highly contagious disease of domestic dogs and some wild carnivores (Budd 1981), and morbidity and mortality may be quite high in susceptible populations (Appel et al. 1981). The disease, caused by a morbillivirus of the paramyxovirus group, occurs throughout the world and is ubiquitous in domestic and wild species (Budd 1981). Closely related viruses include rinderpest and *pest des petite ruminants* of ruminants and measles of man (Appel et al. 1981). The virus, which is fragile in the environment, is readily transmitted by aerosol and by direct contact between infected and susceptible animals. Virus occurs in nasal and ocular discharges as well as urine and feces (Appel et al. 1981). Transmission by fomites is likely when conditions are suitable for virus survival, for example, in an underground burrow where the virus would be protected from sunlight, desiccation, and temperature vacillation. Because canine distemper spreads through susceptible populations primarily by direct contact, transmission should be limited by decreased opportunity for contact caused by declining numbers as animals die or are removed from the population.

All canids, mustelids, procyonids, and at least one viverrid are susceptible to canine distemper (Budd 1981). Domestic ferrets (*Mustela putorius furo*) experience essentially 100 percent morbidity and mortality (Appel et al. 1981, Bernard et al. 1984, Davidson 1986). Susceptibility of the black-footed ferret was demonstrated in the 1970s, when four of six black-footed ferrets vaccinated against canine distemper with modified live virus vaccine died of canine distemper (Carpenter et al. 1976). This vaccine had been previously shown to be safe in domestic ferrets and Siberian polecats (*Mustela eversmanni*). Canine distemper is common in dogs in Wyoming, and wild animals known to be susceptible and present in Park County include coyote (*Canis latrans*), red fox (*Vulpes vulpes*), badger (*Taxidea taxus*), mink (*Mustela vison*), skunk (*Mephitis mephitis*), raccoon (*Procyon lotor*), and weasel (*Mustela frenata*) (Budd 1981, Williams 1982).

The black-footed ferret is the only North American member of the polecat group or subgenus *Putorius* (Anderson et al. 1986). Black-footed ferrets once ranged from the southeastern corner of Alberta and southwest corner of Saskatchewan to Texas (Henderson et al. 1974, Hall 1981). The species seemed nearly extinct until it was rediscovered in September 1981 near Meeteetse, Wyoming (Nowak & Paradiso 1983).

Black-footed ferrets are small, weasel-like animals; females weigh about 700 g and males weigh about 1100 g. Their habitat consists of prairie dog colonies, and they are largely dependent upon prairie dogs for food and shelter. Black-footed ferrets are nocturnal and fossorial. Prairie dog burrows are used for shelter and escape cover; ferrets feed primarily upon prairie dogs, most of which are probably killed underground (Nowak & Paradiso 1983).

The decline of the black-footed ferret has been attributed to conversion of prairie habitat to agriculture and systematic efforts over the past 100 years to eradicate prairie dogs. Ferrets were unintentional victims of prairie dog eradication efforts, and their numbers dwindled as prairie dog towns were plowed under and prairie dogs were poisoned (Anderson et al. 1986, Flath & Clark 1986). A small population of ferrets was discovered in 1964 in South Dakota and studied during the ensuing years. Beginning in 1971, animals from that population were captured for a captive breeding effort by the United States Fish and Wildlife Service (FWS). Only nine animals were available for captive propagation, and four of these were lost to the vaccine-induced canine distemper. Nonetheless, two litters of five each were produced. In each case, four babies were born dead and the fifth died within one or two days despite intensive medical support (Carpenter & Hillman 1979, Hillman & Carpenter 1983, Carpenter 1985). The South Dakota population disappeared in the 1970s.

A Historical Account of Black-footed Ferrets and Disease in Wyoming

A recent and most dramatic example of the importance of disease to endangered species was the 1985 canine distemper epizootic that occurred in Park County, Wyoming, in the only known colony of the black-footed ferrets (Williams et al., in press). Shortly after the black-footed ferret was rediscovered in 1981, the FWS shared with the state of Wyoming responsibility for making decisions necessary for preserving the species. This was done by allowing the Wyoming Game and Fish Department (WGFD) to become the lead agency in management of the black-footed ferret under authority granted to the Service in the Endangered Species Act of 1973. One of the first actions of the WGFD was to form an interagency Black-footed Ferret Advisory Team (BFAT) to advise the department in decision making regarding research and management of the ferret (Carr 1986, Thorne 1987). From the beginning, diseases were recognized by the WGFD and BFAT as a major threat to the ferret colony, and WGFD and BFAT, along with FWS, implemented extensive precautions designed to minimize opportunity for introduction of disease by researchers and managers (Thorne et al. 1985, Thorne 1987).

We are unaware of any free-ranging species, endangered or otherwise, that has received greater consideration about disease. Domestic ferrets, which have been studied extensively, and, presumably, black-footed ferrets are susceptible to many diseases and parasites (Carpenter & Hillman 1979, Bernard et al. 1984). Unfortunately, there is little that can be done to protect a free-ranging mustelid population, no matter how small, from diseases that occur naturally in their environment.

Precautions were taken, however, to minimize the opportunity for introduction of disease by researchers and managers (Thorne et al. 1985). Biologists were not allowed to take dogs to the study site and were admonished to avoid contact with strange dogs, which might have canine distemper. Domestic ferrets are susceptible to human influenza, and, presumably, black-footed ferrets also are susceptible. Consequently, persons suffering from influenza were asked not to participate in field activities, such as trapping, that might bring them into contact with a ferret and lead to its exposure. In addition, everyone near a trapped ferret was required to wear a surgical mask to prevent accidental exposure by someone unaware he or she was infected with influenza. To reduce further the chance of exposing a ferret to influenza or canine distemper, all participants in the process of handling ferrets washed their hands with chlorhexidine scrub (Nolvosan Scrub, Aveco Co., Fort Dodge, Iowa) and rinsed them with virucidal disinfectant (Nolvosan Solution, Aveco Co.). Chemically immobilized ferrets were given a prophylactic dose of penicillin G benzathine and penicillin G procaine, and a mild antibacterial ointment was placed in their eyes. When ferrets were trapped or immobilized for research, a veterinarian was always required to be present (Thorne et al. 1985).

On the advice of a number of consultants, most black-footed ferrets captured for research purposes, which began in 1982, were vaccinated with an inactivated canine distemper vaccine (Williams et al., in press). The same consultants warned that an inactivated vaccine might not be highly immunogenic in black-footed ferrets, that frequent booster doses should be administered, and resistance, if produced, would be of short duration. No attempt was made to trap ferrets for the sole purpose of vaccination because it was not deemed practical.

Concern about the effects of disease was best demonstrated after the discovery, in the spring of 1985, of sylvatic plague in prairie dogs in the complex of towns that supported the black-footed ferrets (Thorne 1987; Forrest et al., in press). This generated a great deal of concern about the future of the prairie dog complex, because it was generally believed that epizootic sylvatic plague results in nearly complete extirpation of prairie dogs in affected colonies (Barnes 1982). Consequently, what may have been the most extensive nonpublic health, nonurban sylvatic plague control effort was enacted; its objective was to preserve the prey base of black-footed ferrets. Approximately 80,000 prairie dog burrows were individually dusted with 5 percent carbaryl to kill flea vectors of sylvatic plague.

As expected, there was no black-footed ferret mortality that could be attributed to sylvatic plague; indeed, subsequent research showed domestic ferrets and Siberian polecats to be resistant to plague (Williams, unpublished data). Although the dusting effort did not eliminate sylvatic plague from the prairie dog complex and there were localized areas where prairie dog mortalities were high, expansion of prairie dogs into other areas resulted in ≤20 percent loss of occupied prairie dog towns by 1986 (Thorne 1987; Forrest et al., in press; Thorne & Belitsky, in press).

In addition to the flea dusting effort, a contingency plan was developed and arrangements made to capture and remove black-footed ferrets from affected prairie dog towns if it was demonstrated that ferrets were being adversely affected by the disease. Plans were made for capture and transport of ferrets to WGFD's Sybille Wildlife Research Unit near Laramie. Caging was available for six animals and temporary caging identified for an additional 20 ferrets. Arrangements were made to move, after a preliminary quarantine period, ferrets in excess of 12 to the Smithsonian Institution's Wildlife Conservation and Research Center, Front Royal, Virginia and the Fish and Wildlife Service's Denver Wildlife Research Center. Because ferrets were not shown to be adversely affected by plague, these contingency plans were not implemented.

The first meeting for the express purpose of discussing captive breeding of black-footed ferrets was held in April 1984 at WGFD headquarters in Cheyenne (Carr 1986, Thorne 1987). That was followed by progressive planning for captive propagation that included a committee report (Anonymous 1984), discussions on captive propagation at BFAT meetings, a workshop in 1984 (Anderson & Inkley 1985), funding requests, a site analysis (Reese 1985), and a recovery meeting at FWS offices in Denver in May 1985 (WGFD 1987). During the May meeting, FWS representatives for the first time expressed optimism that money for captive breeding, according to proposals previously submitted by WGFD, would be available in 1986 or 1987. Consequently, it was decided at that meeting to capture ferrets after field biologists completed 1985 population estimates. The fact that search efforts had failed to locate additional ferret colonies was becoming ominous; the ferrets to be captured in the fall of 1985 were to serve as a hedge against a catastrophe befalling the single known colony of ferrets and also to serve as founder animals for a captive breeding effort (Thorne 1987, Thorne & Belitsky, in press).

During the summer of 1985, preparations were made to house the ferrets in temporary quarters in the existing veterinary laboratory at Sybille. Modifications, including a shower-in facility, were made to protect captive ferrets from exposure to disease and unnecessary disturbance. Three males and three females were to be captured in order to provide the best founder genetic diversity, and one of the males was to be mature to provide an experienced male to enhance the likelihood of breeding. Ferrets were to be captured in pairs at 10- to 14-day intervals to allow time to be sure the first captured animals were healthy and adapting to captivity. Additional ferrets would be captured in 1986 and possibly 1987, and the number taken each year would be small enough so as not to affect the free-ranging population adversely (Thorne 1987; Thorne & Belitsky, in press).

In late September and early October 1985, a scant four years after the colony's discovery, six ferrets were captured and moved to Sybille according to plan. Two adult males were included in the group, and there was a great deal of optimism about the start of captive propagation of black-footed ferrets.

However, the catastrophe the capture of these six animals was supposed to help prevent was already occurring, and they were a part of it. Field observations and population estimates by researchers demonstrated a drastic ongoing decline (Forrest et al., in press) and a meeting was called in Cheyenne on October 22 by WGFD and FWS to discuss possible causes of the decline and courses of action. In addition, the last two ferrets captured were ill with a serious disease they undoubtedly contracted in the wild before capture. One died the day before the Cheyenne meeting and canine distemper was diagnosed (Williams et al., in press).

On October 22, the presence of canine distemper was reported and its probable consequences explained. Although a great deal of precaution had been taken to prevent introduction of disease to captive ferrets from outside the ferret colony, none had been taken to prepare for disease originating from within the colony. Consequently, it was probable that all captive ferrets had been exposed and would die; they did (Williams et al., in press). In addition, trap records showed that two ferrets unwittingly captured with the disease came from widely separated locations, thus indicating that the disease was widespread across the ferret colony. Complete extirpation of the core ferret colony could be expected. The cause of the population decline that apparently had been occurring during the summer and fall could be explained by a canine distemper epizootic.

After much discussion, a decision was made on the evening of October 22 by WGFD and FWS to capture all ferrets that could be located in the main colony; no effort would be made to capture ferrets in more outlying colonies because few had been located there by field biologists during the summer, and a great deal of time and effort would be required in those areas compared to the core colony. The rationale behind the emergency capture was to attempt to remove unexposed ferrets from the colony for future use as founder animals; if left in the colony, it was expected they would ultimately be exposed to distemper and die. If diseased ferrets were captured, their removal from the colony would reduce the opportunity for transmission to other animals. Although very few ferrets had been located in outlying towns during 1985 summer surveys, the fact that none of the 49 animals that were ear-tagged as juveniles in 1984 had been relocated or captured in the core colony in 1985 (WGFD 1987; Forrest et al., in press) suggested that a few may have dispersed into those outlying towns. Low ferret density in these areas justified a hope that a few might have escaped exposure to canine distemper and would repopulate the core area after the epizootic had run its course (Thorne 1987; Williams et al., in press; Thorne & Belitsky, in press).

Six ferrets—two juvenile males, one juvenile female, and three mature females—were captured during the emergency trapping (Thorne 1987). They were placed individually in isolation and none developed canine distemper, indicating they had not been exposed when captured.

Because the two males were apparently physiologically and psychologically immature, no breeding occurred during the 1986 breeding season (DonCarlos et al., in press). During the winter and spring of 1986, WGFD prepared and approved a contingency plan to assist in the decision making process regarding how to manage ferrets that escaped distemper and were located during 1986 summer surveys (WGFD 1986).

Five or six ferrets did escape canine distemper and were located during spotlighting surveys in July and August 1986; two were females that produced litters. Although it was gratifying that a few ferrets remained after the canine distempter epizootic, it was apparent there were not sufficient ferrets present to constitute a viable population. Under guidelines of the contingency plan, a decision was made by WGFD and FWS to capture all remaining ferrets for captive propagation (Thorne 1987, WGFD 1987).

During the summer and fall of 1986 one mature and three juvenile males and two mature and five juvenile females were captured. In February 1987 a mature male was captured, and no sign of another free-ranging ferret has been reported (WGFD 1987).

There are 18 wild black-footed ferrets in captivity and they apparently are all that remain of what is now probably North America's most endangered mammal. The canine distemper epizootic in the only known black-footed ferret colony had devastating effects. The free-ranging colony was essentially extirpated. The captive breeding program quickly went from a carefully planned approach with ideally selected, unrelated founder animals to a crisis situation with related animals, a poor sex ratio, and few mature, experienced breeder males. In addition, there likely will be no opportunity to add to the original 18 animals in the captive colony.

In retrospect, it is unfortunate that ferrets were not captured for captive propagation in 1984 when the population reached its peak size. This had been recommended; certainly WGFD and FWS received much criticism for not having done so (Carr 1986, May 1986). Although temporary quarters could have been prepared at Sybille in 1984 as was done in 1985, in 1984 there was neither promise nor even a suggestion of short- and long-term funding for a captive propagation effort from either federal, state, or private agencies and organizations. Nor had any offers been made by institutions outside of Wyoming to take ferrets and participate in captive breeding.

If black-footed ferrets had been captured in 1984, it likely would have been six to 10 animals. Capture of ferrets in 1984 would not have influenced the 1985 canine distemper outbreak, and most likely it would not have influenced the way WGFD and FWS responded to the epizootic. Consequently, initiation of captive breeding in 1984 would have resulted in 24 to 28 wild caught animals rather than 18. The genetic diversity of the captive population probably would have been better.

There also has been an equal but less articulate criticism that the black-footed ferrets were "studied to death" and their capture led to the population's demise; that they would have been better off it they were never discovered. This, of course, is even less valid. Their nondiscovery would not have prevented the canine distemper epizootic, it simply would have gone unnoticed and the species might now be extinct.

Disease in Other Rare Species

There are other examples of disease severely affecting populations of both free-ranging and captive endangered species. Obviously, epizootics occurring in captivity are more carefully documented. Of the wide variety of methods by which pathogens may be transmitted (Thrusfield 1986), several means seem to pose the greatest danger to captive propagation programs by virtue of detection difficulties or requirement for elaborate management efforts to exclude them. The situation for managers is even more difficult in the field as far less environmental control is possible. Characteristics of the host, especially immunologic status, may also influence the interaction of pathogen and endangered species. A brief discussion of examples of infectious diseases in rare or endangered species follows.

Of particular danger are pathogens brought to endangered animals by the insidious method of carrier animals or reservoir hosts. Carrier animals harbor pathogens and provide sources of infection for endangered species, but they are either unaffected, subclinically affected, incubating the disease and shedding the pathogen, or recovered from the infection but transmitting the disease. The Mauritius pink pigeon (*Columba mayeri*) is threatened with extinction in its native island habitat of Mauritius, and captive propagation is being attempted at several zoos. At the Rio Grande Zoo, Albuquerque, New Mexico, a well-managed, presumably disease-free, flock of domestic pigeons was established because foster rearing appeared to be an ideal solution to a problem encountered when flighty and unpredictable pink pigeons abandoned eggs. However, pink pigeon chicks, hatched by foster pigeons, died suddenly at 5 to 10 days of age. Pigeon herpesvirus was isolated from the dead chicks and was found to be enzootic but undetected in the domestic flock until susceptible pink pigeons were introduced and succumbed to the infection (Snyder et al. 1985). Latently infected pigeons re-excrete herpesvirus during the reproductive period, thus serving as a source of virus for uninfected young (Vindevogel et al. 1985). This case demonstrated the potential hazard from exposure of related but geographically isolated species that differ markedly in their susceptibility to a disease that may be innocuous in one species and fatal in another.

The disease malignant catarrhal fever is transmitted by reservoir hosts to other susceptible species, including endangered mammals. Père David's deer (*Elaphurus davidianus*) exist only in captivity. In 1959 an outbreak of malignant catarrhal fever occurred in the captive herd at Whipsnade Park, United Kingdom, and eight animals in a group of 14 were lost (Huck et al. 1961, Senior et al. 1962). Père David's deer also died in a more recent outbreak in Whipsnade Park (Ashton 1982) and at other zoos (Heuschele 1982). Additional rare and endangered ungulates, including Indian gaur (*Bos gaurus*) (Hatkin 1980, Castro et al. 1982, Heuschele 1982), bongo (*Taurotragus eurycerus isaaci*) (Heuschele 1982), banteng (*Bos javanicus*) (Hatkin 1980, Heuschele 1982), and Arabian oryx (*Oryx leucoryx*) (Heuschele 1982), have also died of malignant catarrhal fever in several zoos. The African malignant catarrhal fever virus (alcelaphine herpesvirus, bovine herpesvirus 3) is well adapted to wildebeest (*Connochaetes* sp.), which carry the virus without apparent disease (Plowright et al. 1960). Virus is shed during parturition and by neonatal wildebeest calves. A similar virus may cause "sheep-associated" malignant catarrhal fever, but the agent has not yet been isolated (Hamdy et al. 1978, Rossiter 1981). Malignant catarrhal fever continues to threaten endangered ruminants held in facilities with reproducing wildebeest and possibly sheep.

In 1978 an outbreak of inclusion body disease of cranes, also caused by a herpesvirus, occurred among captive cranes at the International Crane Foundation, Baraboo, Wisconsin. Some affected cranes belonged to rare species: Stanley (*Anthropoides paradisea*), Manchurian (*Grus aponensis*), and hooded cranes (*G. monacha*) (Docherty & Henning 1980, Docherty & Romaine 1983, Schuh et al. 1986). Thirty-three percent mortality occurred. Carrier birds were probably involved as important sources of the virus. Based on serologic evidence, the virus apparently first occurred in the colony as early as 1975 and remained undetected until overcrowding and environmental conditions in 1978 eventually led to clinical disease.

Canine parvoviral enteritis is a highly contagious infectious disease of canids. Recovered animals may shed the virus for weeks following infection, and the virus is quite resistant to environmental inactivation (Gillespie & Timoney 1981, Kurstak & Tijssen 1981). Deaths among maned wolves (*Chysocyon brachyurus*) due to parvovirus-induced enteritis occurred at the San Antonio Zoological Gardens (Fletcher et al. 1979) and at the National Zoological Park (Mann et al. 1980); carrier wolves or other canids may have been the source of virus for the affected animals.

Infectious diseases transmitted by vectors can be difficult to control in captivity and in the wild without intensive management techniques. In 1984, eastern equine encephalitis, caused by an arbovirus, occurred in a captive flock of endangered whooping cranes (*G. americana*) at the FWS Patuxent Wildlife Research Center, Maryland; seven of 39 birds died (Dein et al. 1986). This widespread mosquito-transmitted viral disease is known to occur naturally or experimentally in many native avian species, but morbidity and mortality is unusual. Sandhill cranes (*G. canadensis*) held at the center were clinically unaffected.

Avian malaria apparently played a major role in the mass extinction of many native Hawaiian birds (Warner 1968, Van Riper III et al. 1986). Introduction of *Culex quinquefasciatus* mosquito vectors in 1826 and *Plasmodium relictum capistranode* protozoans via exotic birds, particularly those from Asia, in the early 1900s was probably required for malaria to have such an impact. The exotic malarian parasite remains and continues to modify distribution and behavior of surviving native avian species (Van Riper III et al. 1986).

Anthropod-borne disease may have unexpected influences on the management of endangered species in ways other than by inducing illness and mortality. Reintroduction of a group of captive raised Arabian oryx into a new national park in Oman was halted due to the presence of serum antibodies to bluetongue virus, an orbivirus, in animals scheduled for release to the wild (Jones 1982). Bluetongue can be an important disease in domestic livestock and is a barrier to movement of animals from bluetongue endemic regions, such as much of North America, to other parts of the world.

Compromise of the immune system would obviously have a major influence on susceptibility of endangered species to infectious diseases. Cheetahs (*Acinonyx jubatus*) are considered threatened. They have been shown to represent an extreme in genetic monomorphism at isozyme and major histocompatibility complex loci that is almost unmatched by other wild species (O'Brien et al. 1983, O'Brien et al. 1985); it has been speculated this genetic uniformity resulted from inbreeding preceded by a major population bottleneck. One of the most successful cheetah captive breeding programs has been at Wildlife Safari, Winston, Oregon, where in 1982 and 1983, 18 cheetahs were lost to feline infectious peritonitis (Pfeifer et al. 1983, Evermann et al. 1983). Rarely does this feline coronavirus-induced disease cause high mortality and morbidity rates in any species that are comparable to those that occurred among Wildlife Safari cheetahs. It has been speculated that the epizootic was so extreme because of the monomorphism of the major histocompatibility complex and lack of effective antiviral immune response (O'Brien et al. 1985). Absence of genetic variability in this species may have robbed it of its natural protection to this coronavirus.

The immune system of animals may also be compromised by prolonged and unavoidable stressors present in the captive or wild environment (Munck et al. 1984). Disease outbreaks in free-ranging and captive animals have been associated with environmental stressors, presumed immunosuppression, and illness due to organisms of low pathogenicity (Fowler et al. 1980, Spraker et al. 1984, Bailey 1986). Prolonged environmental stress increases serum cortisol in captive bighorn sheep (*Ovis canadensis*) (Harlow et al. 1987), which could induce immunosuppression. Physiologic alterations owing to stress may cause fatal hemorrhagic gastroenteritis in some mustelid species (Wallach & Boever 1983; Williams et al., in press).

Iatrogenic diseases may also be important in endangered species; the best examples are induction of disease during immunization against infectious disease using modified live virus vaccines. The canine distemper vaccines are most often incriminated, and vaccine-induced canine distemper has been documented in black-footed ferrets (Carpenter et al. 1976), lesser pandas (*Ailurus fulgens*) (Bush et al. 1976, Itakura et al. 1979), and possibly in African cape hunting dogs (McCormick 1983) and maned wolves (Thomas-Baker 1985).

Humans may also be responsible for transmitting diseases to rare or endangered species. Tuberculosis caused by *Mycobacterium tuberculosis* has created many problems for managers of zoological parks (Montali 1978) and has prevented reintroduction of orangutans (*Pongo pygmaeus*) to the wild in Indonesia (Jones 1982). Measles may cause disease in primates and has caused nearly 100 percent mortality in several groups of colobus monkeys (*Colobus guereza*) (Hime et al. 1975, Mehren et al. 1979). Even species as taxonomically far removed as ferrets may become ill when infected with human influenza virus (Bernard et al. 1984), and this virus could cause disease in black-footed ferrets (Thorne et al. 1985).

Summary

The saga of the black-footed ferret, along with other documented cases of disease in endangered species, provides an example from which lessons may be derived that should be important in the management of other endangered species. Foremost of these is that disease may have catastrophic effects upon a species when it exists in a single population. Timely consideration should be given to dividing a single population, either by translocation or by placing a few animals in captivity.

Precautions were taken to prevent introduction of disease into the free-ranging black-footed ferret population by researchers and managers. We believe these worthwhile and effective and that similar precautions could be tailored to other endangered species. These precautions, however, should not result in complacency, because other animals with which the endangered species shares its habitat are more likely sources of disease.

One precaution taken to protect the black-footed ferret that apprently failed was vaccination against canine distemper. However, this was due to difficulties associated with successfully vaccinating black-footed ferrets against canine distemper (Carpenter et al. 1976; Williams et al., in press). There were no illusions that vaccinated free-ranging black-footed ferrets would be completely protected from canine distemper; rather, it was hoped that exposure to canine distemper virus vaccine would result in some degree of primary immune response and possibly would modify the course of the disease in an individual exposed to the virus. Some endangered species likely can be effectively vaccinated against important diseases, and they should be vaccinated when handled for research or other purposes. Whether or not endangered species should be captured solely for the purpose of vaccination is more contentious, and such a decision should depend upon inherent dangers associated with capture, demonstrated safety and efficacy of the vaccine in question, predicted consequences of the disease, and potential for exposure of members of the endangered species to the disease.

Although isolation and quarantine of individual animals brought into captivity from the same population is generally not considered, had this been done when the first six black-footed ferrets were captured in 1985, there likely would be four additional animals in the captive population. Certainly, isolation and quarantine should be used before recently captured endangered species are formed into a single captive population. During this isolation period all individuals should be screened for as many pathogens of that species as possible.

Addendum

During the 1987 breeding season, nine of 11 black-footed ferret females were bred. The black-footed ferret reproductive cycle was partially characterized; most animals, including juveniles, demonstrated a willingness to participate in breeding activities, and successful pairing techniques were developed. One litter of six kits and one litter of two kits were born; one kit from the litter of two died, and the remaining seven captive-born black-footed ferrets were weaned. Although most breedings were not productive, there is reason for optimism about the future prospects for captive propagation of black-footed ferrets.

Literature Cited

Anderson, E., S. C. Forrest, T. W. Clark, and L. Richardson. 1986. Paleobiology, biogeography, and systematics of the black-footed ferret, *Mustela nigripes* (Audubon and Bachman), 1851. Great Basin Naturalist Memoirs **8**:11–62.

Anderson, S. H., and D. B. Inkley. Black-footed ferret workshop proceedings. Wyoming Game and Fish Department, Cheyenne, Wyoming, USA.

Anonymous. 1984. Black-footed ferret captive breeding facility committee report. Wyoming Game and Fish Department, Cheyenne, Wyoming, USA.

Appel, M. J. G., E. P. J. Gibbs, S. J. Martin, V. ter Meulen, B. K. Rima, J. R. Stephenson, and W. P. Taylor. 1981. Morbillivirus diseases of animals and man. Pages 259–273 *in* E. Kurstak and C. Kurstak, editors. Comparative diagnosis of viral diseases. Vol. 4, Vertebrate animal and related viruses, part B–RNA viruses. Academic Press, New York, New York, USA.

Ashton, D. G. 1982. Malignant catarrhal fever. Veterinary clinical report. Scientific Report, the Zoological Society of London. Journal of Zoology (London) **197**:82–83.

Bailey, J. A. 1986. The increase and die-off of Waterton Canyon bighorn sheep: biology, management and dismanagement. Proceedings of the Northern Wild Sheep and Goat Council **5**:325–340.

Barnes, A. M. 1982. Surveillance and control of bubonic plague in the United States. Symposium of the Zoological Society of London **50**:237–270.

Bernard, S. L., J. R. Gorham, and L. M. Ryland. 1984. Biology and diseases of ferrets. Pages 385–397 *in* J. G. Fox, B. J. Cohen, and F. M. Loew, editors. Laboratory animal medicine. Academic Press, New York, New York, USA.

Budd, J. 1981. Distemper. Pages 31–44 *in* J. W. Davis, L. H. Karstad, and D. O. Trainer, editors. Infectious diseases of wild mammals, 2nd ed. Iowa State University Press, Ames, Iowa, USA.

Bush, M., R. J. Montali, D. Brownstein, A. E. James, and M. J. G. Appel. 1976. Vaccine-induced canine distemper in a lesser panda. Journal of the American Veterinary Medical Association **169**:959–960.

Carpenter J. W. 1985. Captive breeding and management of black-footed ferrets. Pages 12.1–12.13 *in* S. A. Anderson and D. B. Inkley, editors. Black-footed ferret workshop proceedings. Wyoming Game and Fish Department, Cheyenne, Wyoming, USA.

Carpenter, J. W., M. J. G. Appel, R. C. Erickson, and M. N. Novilla. 1976. Fatal vaccine-induced canine distemper virus infection in black-footed ferrets. Journal of the American Veterinary Medical Association **169**:961–964.

Carpenter, J. W., and C. N. Hillman. 1979. Husbandry, reproduction, and veterinary care of captive ferrets. Proceedings American Association Zoo Veterinarians (Washington D.C.) 36–47.

Carr, A., III. Introduction. 1986. Great Basin Naturalist Memoirs **8**:1–7.

Castro, A. E., G. G. Daley, M. A. Zimmer, D. L. Whitenack, and J. Jensen. 1982. Malignant catarrhal fever in an Indian gaur and greater kudu: experimental transmission, isolation, and identification of a herpesvirus. American Journal of Veterinary Research **34**:5–11.

Davidson, M. 1986. Canine distemper virus infection in the domestic ferret. Compendium on Continuing Education for the Practicing Veterinarian **8**:448–453.

Dein, F. J., J. W. Carpenter, G. G. Clark, R. J. Montali, C. L. Crabbs, T. F. Tsai, and D. E. Docherty. 1986. Mortality of captive whooping cranes caused by eastern equine encephalitis virus. Journal of the American Veterinary Medical Association **189**:1006–1010.

Docherty, D. E., and D. J. Henning. 1980. The isolation of a herpesvirus from captive cranes with an inclusion body disease. Avian Diseases **24**:278–283.

Docherty, D. E., and R. I. Romaine. 1983. Inclusion body disease of cranes: A serological follow-up to the 1978 die-off. Avian Diseases **27**:830–835.

DonCarlos, M. W., B. Miller, and E. T. Thorne. Observations of the 1986 black-footed ferret (*Mustela nigripes*) captive breeding program. *In* U. S. Seal, S. A. Anderson, M. Bogan, E. T. Thorne, editors. Reproductive biology of Black-footed Ferrets and small population biology as they relate to conservation. Yale University Press, New Haven, Connecticut, USA, *in press.*

Evermann, J. F., G. Burns, M. E. Roelke, A. J. McKeirnan, A. Greenlee, A. C. Ward, and M. L. Pfeifer. 1983. Diagnostic features of an epizootic of feline infectious peritonitis in captive cheetahs. Proceedings American Association of Veterinary Laboratory Diagnosticians **26**:365–382.

Flath, D. L., and T. W. Clark. 1986. Historic status of black-footed ferret habitat in Montana. Great Basin Naturalist Memoirs **8**:63–71.

Fletcher, K. C., A. K. Eugster, R. E. Schmidt, and G. B. Hubbard. 1979. Parvovirus infection in maned wolves. Journal of the American Veterinary Medical Association **175**:897–900.

Forrest, S. C., D. E. Biggins, L. Richardson, T. W. Clark, T. M. Campbell, III, K. A. Fagerstone, and E. T. Thorne. Population attributes for the black-footed ferret (*Mustela nigripes*) at Meeteetse, Wyoming, 1981–1985. Journal of Mammalogy, *in press.*

Fowler, M. E., N. Jacobsen, L. Erb, and S. McDonald. 1980. Stress-induced mixed infections in Columbian black-tailed deer. Pages 415–420 *in* R. J. Montali, G. Migaki, editors. The comparative pathology of zoo animals. Washington D.C.: Smithsonian Institution Press.

Gillespie, J. H., and J. F. Timoney. 1981. Hagan and Bruner's infectious diseases of domestic animals 7th Ed. Ithaca, NY: Cornell University Press.

Hall, E. R. 1981. The Mammals of North America. New York: John Wiley and Sons.

Hamdy, F. M., A. H. Dardiri, C. Mebus, R. E. Pierson, and D. Johnson. 1978. Etiology of a malignant catarrhal fever outbreak in Minnesota. Proceedings United States Animal Health Association **82**:248–267.

Harlow, H. J., E. T. Thorne, E. S. Williams, E. L. Belden, and W. A. Gern. 1987. Cardiac frequency: a potential predictor of blood cortisol levels during acute and chronic stress exposure in Rocky Mountain bighorn sheep (*Ovis canadensis canadensis*). Canadian Journal of Zoology **65**:2028–2035.

Hatkin, J. 1980. Endemic malignant catarrhal fever at the San Diego Wild Animal Park. Journal of Wildlife Diseases **16**:439–443.

Henderson, R. R., P. F. Springer, and R. Adrian. 1974. The black-footed ferret in South Dakota. South Dakota Department of Game, Fish, and Parks Technical Bulletin no. 4.

Heuschele, W. P. 1982. Malignant catarrhal fever in wild ruminants—a review and current status report. Proceedings of the United States Animal Health Association **86**:552–570.

Hillman, C. N., and J. W. Carpenter. 1983. Breeding biology and behavior of captive black-footed ferrets. International Zoo Yearbook **23**:186–191.

Hime, J. M., I. F. Keymer, and C. J. Baxter. 1975. Measles in recently imported colobus monkeys (*Colobus guereza*). Veterinary Record **97**:392.

Huck, R. A., A. Shand, P. J. Allsop, and A. B. Patterson. 1961. Malignant catarrh of deer. Veterinary Record **73**:457–465.

Itakura, C., K. Nakamura, J. Nakatsuka, and M. Goto. 1979. Distemper in lesser pandas due to administration of a canine distemper live vaccine. Japanese Journal of Veterinary Science **41**:561–566.

Jones, D. M. 1982. Conservation in relation to disease in Africa and Asia. Symposium of the Zoological Society of London **50**:271–285.

Kurstak, E., and P. Tijssen. 1981. Animal parvoviruses: comparative aspects and diagnosis. *In* E. Kurstak and C. Kurstak, editors. Comparative diagnosis of viral diseases. Vol. 3, Vertebrate animal and related viruses, part A—DNA viruses. Academic Press, New York, New York, USA.

Mann, P. C., M. Bush, M. J. G. Appel, B. A. Beehler, and R. J. Montali. 1980. Canine parvovirus infection in South American canids. Journal of the American Veterinary Medical Association **177**:779–783.

May, R. M. 1986. The cautionary tale of the black-footed ferret. Nature **320**:13–14.

McCormick, A. E. 1983. Canine distemper in African cape hunting dogs (*Lycaon pictus*)—possibly vaccine induced. Journal of Zoo Animal Medicine **14**:66–71.

Mehren, K. G., W. A. Rapley, I. K. Barker, S. S. Kalter, and D. Onderka. 1979. Outbreak of a viral disease in newly imported colobus monkeys. Titles and Abstracts, Wildlife Disease Association Conference, 32.

Montali, R. J., editor. 1978. Mycobacterial infections of zoo animals. Smithsonian Institution Press, Washington D.C., USA.

Munck, A., P. M. Guyer, and N. J. Holbrook. 1984. Physiological functions of glucocorticoids in stress and their relation to pharmacological actions. Endocrine Reviews **5**:25–44.

Nowak, R. M., and J. L. Paradiso. 1983. Pages 993–994 *in* Walker's mammals of the world, 4th ed. Vol. 2. Johns Hopkins University Press, Baltimore, Maryland, USA.

O'Brien, S. J., M. E. Roelke, L. Marker, A. Newman, C. A. Winkler, D. Meltzner, L. Colly, J. F. Evermann, M. Bush, and D. E. Wildt. 1985. Genetic basis for species vulnerability in the cheetah. Science **227**:1428–1434.

O'Brien, S. J., D. E. Wildt, D. Goldman, C. R. Merril, and M. Bush. 1983. The cheetah is depauperate in genetic variation. Science **221**:459–462.

Pfeifer, M. L., J. F. Evermann, M. E. Roelke, A. M. Gallina, R. L. Ott, and A. J. Keirnan. 1983. Feline infectious peritonitis in a captive cheetah. Journal of the American Veterinary Medical Association **183**:1317–1319.

Plowright, W., R. D. Ferris, and G. R. Scott. 1960. Blue wildebeest and the aetiologic agent of bovine malignant catarrhal fever. Nature **188**:1167–1169.

Reese, A. 1985. An analysis of potential sites for black-footed ferret captive propagation facility. Wyoming Game and Fish Department, Cheyenne, Wyoming, USA.

Rossiter, P. B. 1981. Antibodies to malignant catarrhal fever virus in sheep sera. Journal of Comparative Pathology **91**:303–311.

Schuh J. C. L., L. Sileo, L. M. Siegfried, and T. M. Yuill. 1986. Inclusion body disease of cranes: comparison of the pathologic findings in cranes with acquired versus experimentally induced disease. Journal of the American Veterinary Medical Association **189**:993–996.

Senior, M., C. R. E. Halnan, and E. H. Tong. 1962. An outbreak of malignant catarrh among the Père David deer. Veterinary Record **74**:932–936.

Snyder, B., J. Thilsted, B. Burgess, and M. Richard. 1985. Pigeon herpesvirus mortalities in foster reared Mauritius pink pigeons. Pages 69–70 *in* Proceedings of the American Association of Zoo Veterinarians. Scottsdale, Arizona, USA.

Spraker, T. R., C. P. Hibler, G. G. Schoonveld, and W. S. Adney. 1984. Pathologic changes and microorganisms found in bighorn sheep during a stress-related die-off. Journal of Wildlife Diseases **20**:319–327.

Thomas-Baker, B. 1985. Vaccination-induced distemper in maned wolves, vaccination-induced corneal opacity in a maned wolf. Page 53 *in* Proceedings of the American Association of Zoo Veterinarians. Scottsdale, Arizona, USA.

Thorne, E. T. 1987. Captive propagation of the black-footed ferret in Wyoming. Pages 419–425 *In* Regional conference proceedings of the American Association of Zoological Parks and Aquariums. Colorado Springs, Colorado, USA.

Thorne, E. T., and D. W. Belitsky. The black-footed ferret captive propagation effort in Wyoming. *In* U. S. Seal, S. H. Anderson, M. Bogan, and E. T. Thorne, editors. Reproductive biology of black-footed ferrets and small population biology as they relate to conservation. Yale University Press, New Haven, Connecticut, USA; in press.

Thorne, E. T., M. H. Schroeder, S. C. Forrest, T. M. Campbell III, L. Richardson, D. E. Biggens, L. R. Hanebury, D. W. Belitsky, and E. S. Williams. 1985. Capture, immobilization, and care of black-footed ferrets for research. Pages 9.1–9.8 *in* S. H. Anderson and D. B. Inkley, editors. Black-footed ferret workshop proceedings. Wyoming Game and Fish Department, Cheyenne, Wyoming, USA.

Thrusfield, M. 1986. Veterinary epidemiology. Butterworth, Stoneham, Massachusetts, USA.

Van Riper III, C., S. G. van Riper, M. L. Goff, and M. Laird. 1986. The epizootiology and ecological significance of malaria in Hawaiian land birds. Ecological Monographs **56**:327–344.

Vindevogel, H., H. Debruyne, and P. P. Pastoret. 1985. Observation of *Pigeon Herpesvirus 1* re-excretion during the reproductive period in conventionally reared homing pigeons. Journal of Comparative Pathology **95**:105–112.

Wallach, J. D., and W. J. Boever. 1983. Diseases of exotic animals. W.B. Saunders Co., Philadelphia, Pennsylvania, USA.

Warner, R. E. 1968. The role of introduced diseases in the extinction of the endemic Hawaiian avifauna. Condor **70**:101–120.

Williams, E. S. 1982. Canine distemper. Pages 10–13 *in* E. T. Thorne, N. Kingston, W. R. Jolley, and R. C. Bergstrom, editors. Diseases of wildlife in Wyoming, 2nd ed. Wyoming Game and Fish Department, Cheyenne, Wyoming, USA.

Williams, E. S., E. T. Thorne, M. J. G. Appel, and D. W. Belitsky. Canine distemper in black-footed ferrets (*Mustela nigripes*) from Wyoming. Journal of Wildlife Diseases, in press.

Wyoming Game and Fish Department. 1986. Recommendations for capture of black-footed ferrets to enhance the captive breeding program in 1986. Wyoming Game and Fish Department, Cheyenne, Wyoming, USA.

Wyoming Game and Fish Department. 1987. A Strategic plan for the management of black-footed ferrets in Wyoming. Wyoming Game and Fish Department, Cheyenne, Wyoming, USA.

Reading 2

Questions and Concerns

READING 2

When you read this article consider the following:

This article examines the state of declining amphibian populations. It also offers several reasons for this decline and that this decline may be an indicator of the planet's health. Do you agree with this statement? What evidence convinced you?

By Beth Livermore

Amphibian alarm: Just where have all the frogs gone?

Habitat destruction does not explain why many species are disappearing in protected areas on five continents at both high altitudes and low

Their eyes bulge, their mouths are ludicrously wide, their fat bellies glisten, their voices are a joke. To some they are synonyms for ugly, but frogs and toads are surprisingly beloved. Mark Twain gave us "The Celebrated Jumping Frog of Calaveras County," Kenneth Grahame wrote Mr. Toad a prominent role in the *The Wind in the Willows,* and for graying adults who grew up on the Smiling Pool stories of Thornton Burgess, a pond is not a pond without the *chug-a-rum* of a bullfrog. It is no accident that they figure prominently in children's stories: How many youngsters have marveled at tadpoles metamorphosing into frogs? How many little boys have come home from a day in the fields with a frog in their pocket? Countryfolk still gig frogs from boats at night, but teenage animal-rights activists increasingly refuse to dissect frogs in their high school biology classes, on the grounds that amphibians are sentient beings, too. Gardeners turn a flowerpot upside down, knock a hole in one side and set it in the dirt, hoping it will become home for a toad that will feast on herbivorous insects. Many more of us equate a frog on its lily pad with the contentment that somehow eludes us.

Now it appears frogs may be fading away, like a number of species of migratory birds. In many parts of the world, a good frog is hard to find. Ponds that once echoed to the crystal plinking of peepers in spring and the tuba riffs of bullfrogs in summer are silent. Nothing plunks into the water as an intruder moves along the shore. Even the biologists whose business it is to know where to find frogs are hard-pressed to locate any. In the past several years herpetologists, the biologists who specialize in amphibians and reptiles, have come to realize that many species are simply disappearing. And no one knows why.

Interviews with biologists on two continents brought the problem home to me. In the Pacific Northwest, I went on a collecting trip with ecologist Andrew Blaustein of Oregon State University at Corvallis. Rather, we tried to collect. Only after half a dozen stops did we finally come up with a single toad, the catch of the day, and that from a roadside puddle. For decades Corvallis scientists have ventured into the nearby Cascade Range to collect amphibians, and it was as easy as falling off a lily pad: frogs and toads lined the shorelines. "We'd go home with dozens after just a few hours," said Blaustein.

About five years ago, however, Blaustein and his colleagues began having to make two and three trips over

Courtesy of Beth Livermore, SMITHSONIAN 10/92

Tombstone Pass to fill their needs. "Now we feel lucky to go home with anything," he said, placing the warty western toad into an empty cooler. "It's weird. There's no obvious reason for this."

Indeed. We had visited several of Blaustein's favorite hunting grounds. Nothing, other than the lack of amphibians, seemed to be wrong with them. Lost Lake, our first stop, looked healthy. Sedges feathered the shoreline. Damselflies, their wings like spun glass, danced among a dazzling array of butterflies. In the center of the lake, which dries to mud come summer, thousands of inch-long toadlets popped up and down.

According to Blaustein there should have been millions. Two years earlier, 50 percent of two million fertilized eggs never hatched. The year before, nearly 100 percent of two million fertilized eggs never hatched. The year before, nearly 100 percent of the million eggs produced lay dead in a gooey white mat. And now only a few thousand had hatched. "We've monitored Lost Lake for 11 years," Blaustein told me. "There has never been more than 5 percent mortality."

The second site was equally idyllic and equally troubling. Lily pads and silver logs sat devoid of web-toed feet. Nothing moved. "Two years in a row Cascades frogs laid eggs here; they hatched into tadpoles, they did fine for a few weeks, and then they completely disappeared," recalled Blaustein, stirring the water idly with an empty net. The site has been studied since 1975. "Nothing like this has ever happened before." Only after half a dozen more stops did we find our toad. "That's not just inconvenient, it's scary. The massive egg mortality and the disappearance of larvae are unprecedented in Oregon." Now Blaustein looks for clues. Like hundreds of other biologists around the globe, he thinks that ailing amphibian populations may reflect the sickly state of their environments — and our world.

Blaustein is not alone. At the first World Congress of Herpetology, held in 1989 in Canterbury, England, scientists from all over the world swapped similar stories and concerns. David Bradford of the University of California said yellow-legged frogs had vanished from 37 of the 38 ponds where he had studied them ten years earlier. Michael Tyler from the University of Adelaide in Australia had not seen a gastric brooding frog, which raises its young in its stomach, since 1980. Martha Crump of the University of Florida in Gainsesville said the golden toad, native to Costa Rica's Monteverde Cloud Forest Reserve, had virtually disappeared.

Alarm gave rise to an emergency workshop the following February in Irvine, California. More stories surfaced: frogs, toads and other amphibians have vanished from parts of Denmark, Nova Scotia, Peru and Panama. Places as different as the Swiss Alps and the wetlands of India are missing species. In all, 16 countries, representing five continents, reported massive dips and die-offs among their amphibians. The implications are profound.

Amphibians are among the oldest creatures on Earth. They evolved more than 350 million years ago and shared their steamy swamps with dinosaurs. "If they disappear now, after all of that time, there's something to be concerned about," says biologist David Wake, director of the Museum of Vertebrate Zoology at the University of California at Berkeley. In addition, amphibians are a vital link in the food chain: larvae eat plants and/or animals, and predatory adults consume a wide spectrum of creatures; reptiles, birds, fish and mammals, in turn, eat them.

Moreover, it may be that amphibians signal environmental stress earlier than do most other organisms. Most have dual habitats, thereby risking exposure to both water-based and land-based contaminants and predators. They have permeable skin that interacts with water, air and dirt. They lack feathers and hair, so small increases in ultraviolet radiation may cause immediate damage. Many biologists say that amphibians are the canaries in the global coal mine.

Herein lies the debate. Cautious observers point out that much of the evidence is anecdotal. "That's not the kind of evidence that a quantitative ecologist would feel comfortable with," says biologist Peter Morin, of Rutgers University in New Jersey. Furthermore, critics assert that little is known about the biology and behavior of most species. So, determining whether an extinction is natural or induced by mankind is, at best, difficult. In addition, what may look like an extinction could instead be a temporary absence because of hibernation or dormancy. "You'd need about a hundred years of data to know what is 'normal,'" says Joseph H. K. Pechmann, coauthor of a 12-year study at the University of Georgia's Savannah River Ecology Laboratory in South Carolina. His investigations show wild fluctuations but overall stability.

Some species leapfrog the trend

Adding to the uncertainty, amphibians in some areas are doing quite well. Penn State biologist Blair Hedges, who has worked in the West Indies for the past ten years, reports: "Within the past two years we have heard or observed all 28 species of frog ever known to have occurred at those sites. In fact, we just added two species." The cane toad has overrun southeastern Queensland in Australia (SMITHSONIAN, October 1990). And then there are places where both vanishing and stable species live side by side. For example, thriving wood frog and declining northern leopard frog colonies reside in the same area of the Rocky Mountains.

Still, even skeptics believe that the question of decline

is worth investigating. "I'm not saying that people shouldn't study the problem. I'm saying they should do it rigorously and using a protocol that you can apply from one place to the next so that you can compare results," says Morin. Others call for immediate action. "If you wait for the final word — definitive research — there may not be anything left to study," says Russell Mittermeier, president of Conservation International in Washington, D.C. "When it comes to problems of this magnitude, I'm in favor of jumping to conclusions." A number of herpetologists, a generally conservative bunch, appear to agree with him.

In December 1990 the International Union for the Conservation of Nature and Natural Resources (IUCN) assigned its Species Survival Commission to create a task force to explore the problem. "Out of this request came the Declining Amphibian Populations Task Force," says James Vial, the international coordinator. "We're applying a SWAT team approach to the problem. We want to get in there, figure out what's going on and get out as fast as possible."

His office is located in the Environmental Protection Agency's research laboratory in Corvallis — just down the road from the Blaustein team. "But this is an international group," he says, diagraming the family tree on a chalkboard. The world will be divided into working groups. He has already assigned chiefs of staff to more than 40 groups in 30 countries. Each group will evaluate the health of local populations and probe peculiarities. He will enter the resulting information into a "Frog Log" database, which will be available to scientists worldwide.

The Australian work group sent 600,000 questionnaires to its citizenry for population data. Scientists working in Panama and Costa Rica have met to coordinate their work and to prepare a proposal for future joint studies. A team at the Smithsonian's National Museum of Natural History is developing a protocol for worldwide data collection. "It will be a complete, self-contained manual for scientists, fish-and-game people, environmental consultants — anyone interested in quantifying amphibian populations," says coauthor Roy McDiarmid. He hopes it will lead to more reliable data.

Biologists are looking for probable causes as well as measuring declines. When the European Herpetological Society (SEH) met in Budapest last fall, an IUCN committee added another 14 species to an already crowded list of threatened frogs and toads. "Habitat destruction and landscape planning are the major threats to these species," the SEH said. Most of the damage in Europe was done long ago and is not as visible today as it is in Central and South America, where daily destruction of rain forest may be wiping out frogs faster than they can be identified. But some is recent. "Before World War II, white storks sat on every rooftop in many German towns," said Ulrich Joger, an SEH officer. After the war, development swept through Germany. "Now you never see them anymore," he said wistfully. "They feed mainly on frogs."

In some areas the German government has dug new ponds and transferred populations of frogs and toads. But it also stocks them with fish, and all too often the frogs are soon gone. French biologist Alain Dubois told me that fish planting was also common in the natural lakes of Europe. He had discovered a subspecies of newt in what was then Yugoslavia: "There were thousands of them in 1978." A few years later he sent a student to collect samples. No more newts. Instead, the student found introduced trout. In France goldfish gobble frog eggs from ponds and lakes, he said. "People bring them home from holiday and don't know what to do with them, so they dump them into the nearest pond. Zap! The frogs are gone."

Dubois is assistant director of the Musée National d'Histoire Naturelle in Paris and general secretary of the International Society for the Study and Conservation of Amphibians. I had to ask him how much the French passion for grenouille — frogs' legs — was affecting the decline of frog populations. "It may contribute to the problem," he conceded. It was in the 1950s that demand began to exceed the local supply of the species *Rana esculenta* (the common name is "edible frog," even though most frogs are edible). In the '90s the majority of frogs' legs sold in France are those of Asian bullfrogs imported from Indonesia and Bangladesh. "Today france consumes 3,000 to 4,000 [metric] tons of frogs' legs each year," said Dubois. "That's about ten times more than 20 years ago." (It takes at least 20,000 frogs to make up a metric ton.)

These days the Asian bullfrog is becoming increasingly more difficult to find. Some scientists suspect that many of the so-called Bangladesh frogs are actually being smuggled from India, where exports were banned in 1987 because insects were getting out of control. India has been forced to introduce a new frog species and increase the use of insecticides.

Dubois leaned toward me. "Do you want to discuss anything else?" he asked, a little annoyed. "There are a lot of other things to consider," he said. Indeed, predation does not explain the mysterious die-off of clutches of frog eggs in the protected Fountainbleau forest, south of Paris. In fact, the most confounding disappearances of amphibians are those occurring in preserves and parks worldwide, where hunting, fishing and farming are prohibited. Some scientists believe the most important factors in this mystery are still to be unveiled. Among the favored hypotheses: acid rain, global warming, pathogens and increased ultraviolet radiation.

To learn about the first of these, I headed to Penn State and a symposium titled "Habitat Acidity and Amphibian

Decline." A joint meeting of the Society for the Study of Amphibians and Reptiles and the Herpetologists' League heard eight hours of papers that appeared to indicate that acid deposition (by rain or snow or in dry form) hurt many, but not all, amphibians. Frogs and toads are most vulnerable early in life. One study showed that their eggs, when bathed in slightly acidic water, either die or produce deformed tadpoles. Several others showed that increased acidity can slow growth rates in frogs, which means trouble for species that breed in temporary ponds and must mature before the water evaporates. Still other evidence indicated that highly acidic water kills frog sperm.

There are exceptions, however. Some Eastern amphibians, such as the New Jersey Pine Barrens frog, have long been exposed to acidic environments and show little effect. One explanation proposed at the symposium: genetic adaptation. Scientists theorize that these frogs may have evolved in their boggy, acidic habitat to avoid competition. Swedish biologist Claes Andren, of the University of Goteborg, presented evidence that a population of moor frogs exposed to high acidity for approximately 15 generations showed increased adaptation to acidic conditions. But this sort of adaptation does not help frog populations that are suddenly exposed to greater acidity.

Some scientists argued that acid rain is not a reasonable explanation for frogs' disappearance at high altitudes because precipitation is less acid there. Melting snowbanks, however, can produce a "pulse" of accumulated acidity in just a few hours. And this pulse flushes into lakes and ponds in the spring just as frogs, toads and salamanders are mating — the most crucial time in their life cycle.

The slightest warming of the globe could have a major effect on amphibians. Many scientists believe that any warming will bring changes in precipitation patterns as well. And even without global warming, of course, there are wide fluctuations in precipitation from year to year and decade to decade. Even small differences in the amount or timing can spell doom for animals that depend on water for survival and reproduction. Amphibians need a reliable water source because their permeable skins leak body fluids rapidly and need to be kept moist for proper respiration. Frogs and toads must have water to lay and fertilize their eggs; tadpoles need water for development and growth.

An example of weather change as disaster was offered by Richard Wyman, director of the Edmund Niles Huyck Preserve and Biological Research Station in Rensselaerville, New York. In 1988 he studied breeding patterns of the yellow-spotted salamander in south central New York. Traditionally, males journey to ponds during the first spring rain, just after the ground defrosts. Females migrate the next rainy day. That year, 23 days of dry weather followed the male pilgrimage. When the females finally made their way to the pond, most of the males had already moved on.

"Five of the 12 driest months on record and six of the ten warmest years since 1855 have occurred during the 1980s," said Wyman. Climatologists expect global warming to cause continental interiors to be drier, and droughts more intense. There is also some evidence that another predicted effect of global warming is happening now: the climatic and ecological conditions to which animals and plants have adapted may be moving northward at 300 to 600 miles per century.

The last of the golden toads?

On the telephone to Costs Rica I learned that biologist Alan Pounds, who lives and works in the Monteverde Cloud Forest Reserve, thinks a disruption of the water cycle probably played a role in the sudden disappearance of the golden toad and other species in his area. In 1988 the University of Florida's Martha Crump spotted just one golden toad. Last spring Pounds found no golden toads, no yellow-and-black harlequin frogs and just a few lime-green glass frogs. "Everything seems less abundant," he said.

Between the 1987 and 1988 breeding seasons, Pounds says, the number of golden toads and harlequin frogs observed declined by 99 percent. He notes that there is evidence that the population crashes of these two species may be linked to the unusually warm, dry conditions in 1986 and 1987. These climatic effects, combined with changes in groundwater levels caused by deforestation outside the Monteverde Reserve, may have been too great for the animals to withstand. Pounds adds that the last time harlequin frogs were censused in the 1980s, they were stressed from lack of moisture. Thus he does not rule out the idea of pathogens possibly having affected the stressed animals.

In Oregon, Andrew Blaustein had talked about the third general explanation for the apparent worldwide drop in amphibian populations: an increase in the ultraviolet (UV) radiation reaching Earth's surface. The flux has been increasing as the ozone layer 30 miles up has thinned, not only over Antarctica but, to a lesser degree, everywhere. This kind of exposure harms plants and animals, including humans. In Antarctica, biologists found that UV radiation damages the DNA of some phytoplankton, microscopic plants growing in the ocean. In 1975 an Oregon State University biologist had discovered that toad eggs and tadpoles (his sample had come from Lost Lake) are very sensitive to increased radiation in the middle of the ultraviolet range, commonly referred to as UV-B. (This is the same range that does the most serious damage to human skin.) Reports of frog loss seem to come more often from higher altitudes and latitudes than from lower altitudes and the tropics.

Blaustein thinks the correlation is worth investigating. So in spring of the past two years he filled a plastic jug with frog eggs from Lost Lake before the die-offs and reared them in his laboratory, away from direct sunlight. The tadpoles, raised in water from the lake, developed normally. Blaustein and his team are currently surveying 70 historically rich frog ponds to document their overall population status. He is preparing to conduct laboratory and field experiments to see how various levels of UV affect developing amphibian larvae, reproductive endocrinology and adult behavior in amphibians.

A colleague of Blaustein's, John B. Hays, has been trying to find whether frog and toad eggs can repair damage to their DNA caused by high levels of UV radiation. For nearly a year, Blaustein has delivered wild amphibians to Hays, an agricultural chemist, who removes eggs and oocytes, precursors to eggs, from them. "With those extracts we can measure two different kinds of DNA repair activity," said Hays. "The standard lab frog, *Xenopus laevis,* or African clawed frog, has a tremendous capacity to repair UV damage." If the same holds true for their wild cousins, UV may be acquitted of local egg clutch die-offs.

I left this world of test tubes and microscopes to say goodbye to the toad that Blaustein and I had caught for Hays earlier in the day. It would be killed as part of his research. Many a frog and toad have died for human benefit. Indians used the skin secretions of poison-arrow frogs to tip their darts. Frogs' legs gave us an understanding of electrical nerve impulses as well as gourmet dining. Those African clawed frogs have recently offered up clues to cystic fibrosis and Parkinson's disease. But as I walked down the hall, I found myself thinking not of their utility but of how much poorer the world would be without those unhandsome guardians of pond and garden. I made my goodbyes, shut the cooler top and headed for home.

Questions and Concerns

READING 3

When you read this article consider the following:

This paper presents the results of a study conducted by several governmental agencies concerned with forest and wildlife management. Do you feel that the methods used to study the spotted owl were adequate?

Biological Conservation 1992, **62**, 95–100

Reading 3

Dependence of northern spotted owls *Strix occidentalis caurina* on old-growth forests in the western USA

Jonathan Bart

Ohio Cooperative Fish and Wildlife Research Unit, US Fish and Wildlife Service, and Department of Zoology, Ohio State University, 1735 Neil Avenue, Columbus, Ohio 43210, USA

&

Eric D. Forsman

Forestry Sciences Laboratory, US Forest Service, Olympia, Washington 98502, USA

(Received 4 March 1991; revised version received 18 October 1991; accepted 4 November 1991)

Habitat requirements of northern spotted owls *Strix occidentalis caurina* have become the focus of a major controversy over how much old-growth forest in the western United States should be preserved. Analysis of three large data sets showed that the subspecies was rare or absent in areas with little older (i.e. >80-year-old) forest but with extensive stands nearing harvest age. The owls were also rare in areas with the small amounts of old-growth typically left after harvest operations. Old-growth stands in Wilderness Areas supported sparse populations of northern spotted owls, and their reproductive success was only about half that of owls outside Wilderness Areas. The results indicate that timber harvest operations, as currently practiced, lead to declines in numbers of northern spotted owls, and that currently protected old-growth stands are unlikely to provide enough high-quality habitat to maintain self-supporting populations of northern spotted owls.

INTRODUCTION

The northern spotted owl *Strix occidentalis caurina* has recently become the focus of an intense debate over how the remaining old-growth and mature forests on public lands in the western United States should be managed. Although the controversy raises pervasive questions about how the American public wishes to allocate resources between timber production and forest preservation, the specific issue being examined most closely is how dependent spotted owls are on old-growth forests.

Old-growth forests within the range of the northern spotted owl now occur mainly on public land (Thomas *et al.*, 1990). If harvest rates during the past few decades continue, most of the currently unprotected old-growth in the range of the northern spotted owl will disappear within a few decades (Fig. 1). Decisions about the dependence of this subspecies on old-growth forests thus need to be made during the next few years or the options for protecting the remaining old-growth forests will be sharply curtailed.

The US Fish and Wildlife Service has recently listed the northern spotted owl as Threatened under the Threatened and Endangered Species Act. This new designation for the subspecies, which formerly had no special protection under federal law, is likely to have profound effects on how much old-growth is harvested on public land within its range (much of western Washington, Oregon, and California). Still to be decided, however, is the difficult question of just how much older forest is needed by the owls.

The dependence of northern spotted owls on old-growth forest has been hotly debated during

Biological Conservation 0006-3207/92/$05.00

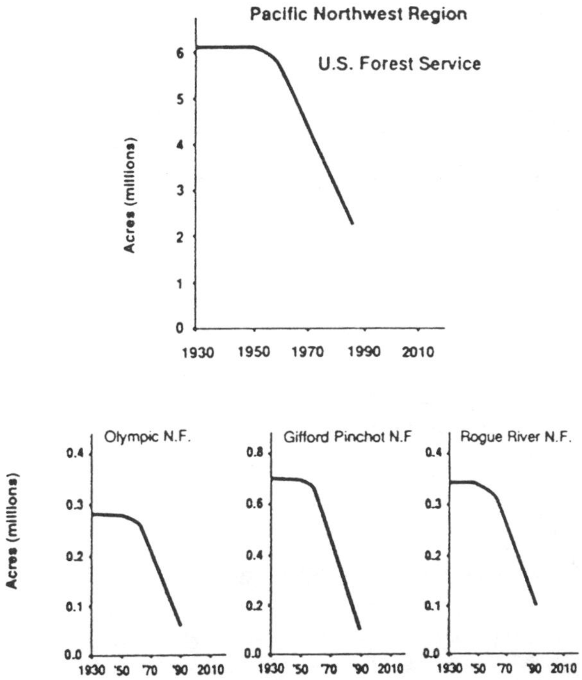

Fig. 1. Decline of northern spotted owl habitat on US Forest Service land in Washington and Oregon outside of Wilderness Areas (from Anderson *et al.*, 1990).

the past several years. Many biologists contend that the subspecies is largely dependent on forests more than 80 years old (referred to in this paper as 'older forest') (Thomas *et al.*, 1990). They point out that few owls have been found in extensive areas lacking older forests (Forsman *et al.*, 1977), and telemetry studies show that birds concentrate their activities in older forest and avoid very young forests (Forsman *et al.*, 1984). Critics, however, have argued that few thorough searches have been reported from areas where regenerating forests are near harvesting age (expected to be about 80 years—Brown, 1985), and that the telemetry studies do not prove that spotted owls require older forest, but only show that they select older forest when it is available (Irwin, 1989). Critics also note that recent surveys in California revealed spotted owls in forests 50–80 years old, and the birds there appeared to be reproducing successfully (Irwin *et al.*, 1989). Thus, the debate over how dependent spotted owls are on older forests has been difficult to resolve conclusively because too few studies, surveying both older and younger forests with similar methods, have been published.

During the past two to three years, this situation has changed dramatically as the US Forest Service, Bureau of Land Management, and State Wildlife Departments have begun extensive surveys for spotted owls on the lands they manage. Here, we use the results from these surveys to address three questions, all of which bear on the larger issue of how timber harvest activities are likely to affect spotted owls: (1) Do landscapes consisting almost exclusively of young (i.e. <80-year-old), regenerating forests support northern spotted owls? (2) Do landscapes with the small amounts of older forest left after timber harvest operations (usually <20%, Anderson *et al.*, 1990) support northern spotted owls? (3) Do protected forests in National Parks and Wilderness Areas provide high-quality habitat for northern spotted owls? These protected areas are widely distributed and could, if they contain high-quality habitat, provide an extensive system of refuges for the owls. Wilderness Areas and National Parks, however, are often placed on high-elevation land, whereas northern spotted owls are thought to reach higher abundance at lower elevations (Thomas *et al.*, 1990). There is thus some question about the suitability of the habitat in these protected sites.

METHODS

We used results of surveys for northern spotted owls on areas scheduled for timber harvest, on intensively studied research areas, and on randomly selected, 400-ha circles on US Forest Service lands. For the first two data sets, we obtained habitat maps of each surveyed area and delineated compartments within which the distribution of older forest was fairly uniform. We then determined the fraction of each compartment covered by older forest. We restricted the analysis to compartments of at least 20 km² to minimize edge effects. Most timber sale areas were surveyed three times in 1989, though a few were surveyed more times or also in 1988. Intensively surveyed areas were visited up to 12 times per year, most for at least two consecutive years. Results from these two data sets were very similar (Anderson *et al.*, 1990) and are therefore combined here.

Two sets of 400-ha circles were available. The first was randomly selected from US Forest Service land throughout the range of the northern spotted owl. The second was randomly selected from 'spotted owl habitat areas' (SOHAs) on US Forest Service land in Washington and Oregon. Most circles were surveyed in 1988 and 1989, and all were surveyed following a detailed protocol requiring up to six visits (O'Halloran, 1989; Simon-Jackson, 1989). Surveyors recorded number of owls, number of pairs, and, in many cases, number of young fledged per pair.

We investigated the first issue—owl abundance in young, regenerating forests—by selecting compartments with extensive stands 50–80 years old, and little older forest. Each was paired with a nearby compartment having extensive older forest. We compared the density of owls, and pairs of owls, in these two habitats.

Owl abundance and productivity on areas with <20% older forest was investigated by calculating rangewide estimates of the average number of owls/km², pairs/km², fledglings/km², and fledglings/pair in compartments with different amounts of older forest. Our main purpose was to compare compartments with <20% and >60% older forest, but we also report the measures above for compartments with intermediate levels.

The third issue—whether currently protected older forest provides high quality habitat for northern spotted owls—was studied by two analyses of survey results from the randomly selected, 400-ha circles. The first used the circles selected from throughout the range of northern spotted owls on US Forest Service land. We compared number of owls/km², pairs/km², fledglings/km², and fledglings/pair inside and outside Wilderness Areas in good habitat (defined as circles with >40% older forest). The second analysis, carried out using circles in SOHAs, was similar except that we only compared young fledged per pair. Measures of abundance were not compared because SOHAs were usually placed in locations known to contain spotted owls.

RESULTS

Information was obtained from 185 compartments, including five intensive sites, in 45 study areas (Fig. 2). The compartments varied in size from 20 to 690 km², and covered a total area of 11 057 km². Seven compartments, covering a total area of 1029 km², had extensive 50–80-year-old forest but little forest >80 years old. Data were available from 254 of the 400-ha circles. The only physiographic province not covered by our data was the redwood zone in coastal California, which covers less than 7% of the range of the northern spotted owl (Anderson *et al.*, 1990).

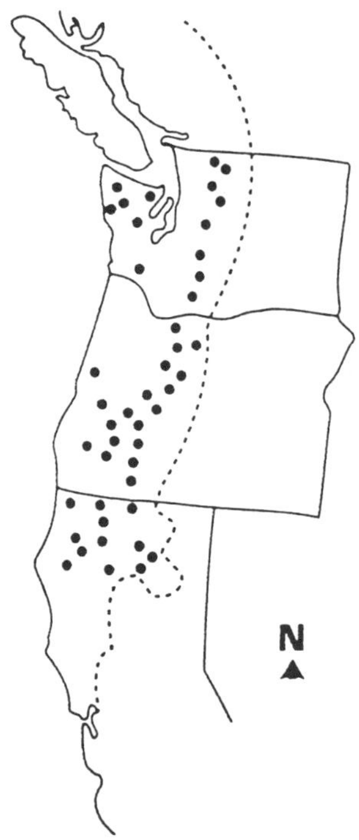

Fig. 2. Location of study areas in Washington, Oregon and California. All areas on public land, of at least 164 km², that had been surveyed at least three times in at least one year were included in the analysis.

Virtually no owls were found in compartments dominated by 50–80-year-old forest but lacking older forest (Table 1). In contrast, detections of single and paired owls were common in the nearby compartments with extensive older forest. Owls were approximately 40 times more common on the areas with older forest than on the areas lacking older forest. This analysis indicates that

Table 1. Density of northern spotted owls in regenerating stands 50–80 years old and in nearby areas with extensive older forest

State	No. of sites	Total area (km²)	Percentage of area covered by 50–80-year-old stands	Owls per km²	Pairs per km²	Owls/km² in adjacent sites with extensive older forest
Washington	4	207	50–92	0·00	0·00	0·05
Oregon	2	417	57–68	0·02	0·01	0·32
California	1	474	59	0·00	0·00	0·27

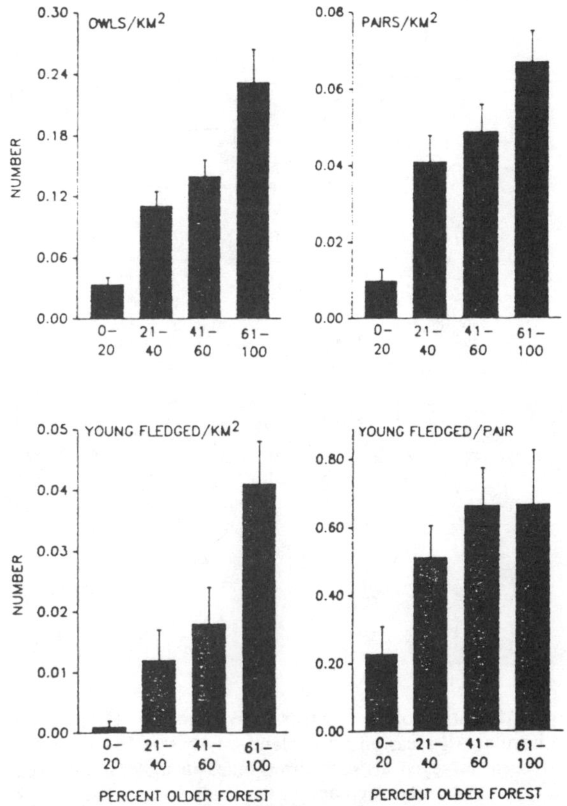

Fig. 3. Density and reproductive success of northern spotted owls in relation to amount of older forest on the surveyed area. Vertical lines indicate 1 SE.

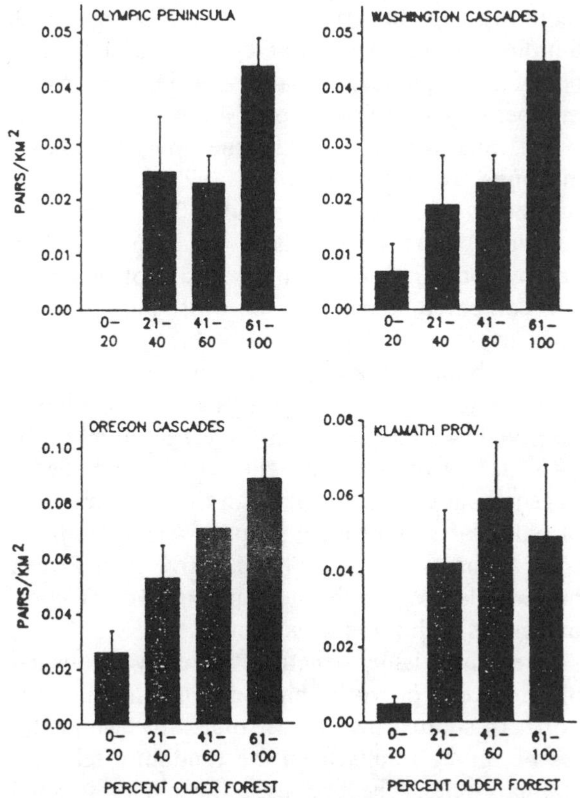

Fig. 4. Abundance of spotted owls, in four portions of their range, in relation to the amount of older forest on the surveyed areas. Vertical lines indicate 1 SE.

50–80-year-old forests do not provide suitable habitat for northern spotted owls.

Surveys on timber sale areas and intensively studied areas (Fig. 3) showed that abundance and productivity increased steadily with increasing amounts of older forest. All measures of abundance or productivity were significantly lower ($p <$ 0·05, t test) on areas with <20% older forest than on areas with >60% older forest. Approximately 50 times more young owls fledged/km² in areas with >60% older forest than in areas with ≤20% older forest. Similar results were obtained from the 400-ha circles (Table 2). Trends were similar in different parts of the range (Fig. 4) and in each of the five intensively studied areas (see Anderson *et al.*, 1990 for additional details). This analysis shows that landscapes with less than 20% coverage by older forest rarely provide suitable habitat for northern spotted owls.

Table 2. Density and reproductive success of northern spotted owls in relation to amount of older forest as indicated by US Forest Service monitoring data

Variable	Percentage suitable habitat			
	0–20	21–40	41–60	> 60
Owls/site	0·31^{a}	0·64^{b}	0·85^{b}	0·95^{b}
Pairs/site	0·04^{a}	0·19^{b}	0·29^{b}	0·28^{b}
Fledglings/site	0·01^{a}	0·13^{b}	0·21^{b}	0·22^{b}
Fledglings/pair	0·33	0·77	0·67	0·93
Number of sites	101	56	58	39

Different superscripts within rows indicate significantly different values (t test, $p < 0{\cdot}05$).

Table 3. Density and reproductive success of northern spotted owls in randomly selected sites within Wilderness Areas and in good habitat (older forests covering >40% of the site) outside of Wilderness Areas

Variable	Location of site	
	Wilderness	Good habitat
Owls/site	0·61^{a}	0·97^{b}
Pairs/site	0·18^{a}	0·29^{b}
Fledglings/site	0·14^{a}	0·43^{b}
Fledglings/pair	0·54^{a}	0·96^{b}
Number of sites	104	51

Different superscripts within rows indicate significantly different values (t test, $p < 0{\cdot}05$).

All measures of abundance and productivity in the 400-ha circles were lower inside than outside Wilderness Areas (Table 3). Productivity inside Wilderness Areas was only 30–50% as high as productivity outside. Similar results were obtained from circles in SOHAs. Pairs inside Wilderness Areas fledged an average of 0·42 young per attempt whereas pairs outside Wilderness Areas produced 0·89 fledglings/year ($p < 0{\cdot}05$, t test). This analysis shows that land within Wilderness Areas classified as spotted owl habitat does not, on average, provide high-quality habitat for the owls.

DISCUSSION

This study showed that northern spotted owls are rare, and have low productivity, in 50–80-year-old forests or areas with only the small amounts of older forest left after timber harvest activities. Owls were virtually absent in 50–80-year-old forests (Table 1). The study areas covered >1000 km² and were surveyed for two years, but only one pair (in one year) was found and it did not produce any fledglings. On areas with <20% older forest a few single owls were found but pairs were scarce, and productivity was far below the levels found on areas with extensive older forest. This pattern was found on timber sale areas and intensively studied areas (Fig. 2), on randomly selected, 400-ha circles (Table 2), and in four different portions of the range (Fig. 3). The general pattern appears to be that a few solitary owls use areas with <20% older forest but that these areas only rarely provide habitat of sufficient quality to support successful reproduction.

The study also showed that the older forest in Wilderness Areas does not generally provide high-quality habitat for northern spotted owls. Abundance and productivity were somewhat higher than in areas where the timber had been harvested, but were still significantly below the levels found in high-quality habitat outside Wilderness Areas. No comparable survey data were available from National Parks, but much of the older forest in these areas, like that in the Wilderness Areas, is at high elevation and seems unlikely to provide substantially better habitat than the older forest in Wilderness Areas. Thus, while surveying older forest in National Parks should be a high priority, it appears that currently protected older forest may be unable to support self-sustaining populations of northern spotted owls.

The issue of whether protected land will be adequate for protecting spotted owls has previously been addressed with demographic models (e.g. Lande, 1988; Doak, 1989). Some investigators have reached different conclusions, and obtaining a consensus has not been easy (Anderson *et al.*, 1990; Thomas *et al.*, 1990). Our comparison of abundance and productivity in protected and unprotected areas represents a different approach, perhaps most closely aligned with the 'source–sink' view of population dynamics (Pulliam, 1988). This paradigm applies well to northern spotted owls residing inside and outside of the protected areas. Owls in protected areas are less abundant than in good habitat outside protected areas, they are at the upper edge of their elevational range, and, most importantly, they exhibit sharply reduced productivity compared to populations in good habitat. These statements are true both of the protected areas in general and of the best sites (the SOHAs) within the protected areas. We suspect that the source–sink model may be useful for evaluating management strategies in many situations where only a small amount of the existing habitat is guaranteed protection.

ACKNOWLEDGEMENTS

Numerous federal, state, and private wildlife biologists helped us to obtain the data used in this study. We thank the following for advice during our analyses: D. R. Anderson, T. A. Bookhout, S. Earnst, T. C. Edwards, Jr, A. B. Franklin, G. Gunderson, R. Gutierrez, C. B. Kepler, J. B. Lint, E. C. Meslow, B. R. Noon, and J. Verner. This research was supported by the US Fish and Wildlife Service and the US Forest Service.

REFERENCES

Anderson, D. R., Bart, J., Edwards, T. C. Jr, Kepler, C. B. & Meslow, E. C. (1990). *1990 Status Review: Northern Spotted Owl*. US Fish and Wildlife Service, Washington, DC.

Brown, E. R. (1985). Management of wildlife and fish habitat in forests in western Oregon and Washington. *US For. Serv. Tech. Rep.*, No. RG-FGWL-192-1985, Portland, Oregon.

Doak, D. (1989). Spotted owls and old growth logging in the Pacific Northwest. *Conserv. Biol.*, **3**, 389–96.

Forsman, E. D., Meslow, E. C. & Strub, M. J. (1977). Spotted owl abundance in young versus old-growth forest, Oregon. *Wildl. Soc. Bull.*, **5**, 43–7.

Forsman, E. D., Meslow, E. C. & Wight, H. M. (1984). Distribution and biology of the spotted owl in Oregon. *Wild. Monogr.*, **87**.

Irwin, L. L. (1989). *Technical Review of 'The Northern Spotted Owl: a Status Review Supplement'*. National Council of the Paper Industry for Air and Stream Improvement, Corvallis, Oregon.

Irwin, L. L., Self, S. & Smith, L. (1989). Status of the Northern Spotted Owl on Managed Forestlands in Northern California. National Council of the Paper Industry for Air and Stream Improvement, Corvallis, Oregon.

Lande, R. (1988). Demographic models of the northern spotted owl *Strix occidentalis caurina*. *Oecologia, Berl.*, **75**, 601–7.

O'Halloran, K. (1989). *Spotted Owl Inventorying and Monitoring: Annual Report for 1989*. US Forest Service, Pacific Northwest Region, Portland, Oregon.

Pulliam, H. R. (1988). Sources, sinks, and population regulation. *Amer. Nat.*, **132**, 652–61.

Simon-Jackson, T. (1989). *Spotted Owl Inventory and Monitoring Program: Annual Report for 1989*. US Forest Service, Pacific Southwest Region, San Francisco, California.

Thomas, J. W., Forsman, E. D., Lint, J. B., Meslow, E. C., Noon, B. R. & Verner, J. (1990). *A Conservation Strategy for the Northern Spotted Owl*. Interagency Scientific Committee to Address the Conservation of the Northern Spotted Owl, US Government Printing Office, Washington, DC.

Questions and Concerns

READING 4

When you read this article consider the following:

This article draws attention to the spotted owl issue from the loggers point of view. While it focuses on how the people of these areas are being adversly affected, it still states that the "remaining unprotected old growth on the Peninsula will be logged within ten to fifteen years at pre-1990 harvest rates." What governmental department does the author of this article blame for the current controversy?

Timber Towns in Trouble

"We're living here in a small timber community, a very patriotic community," says Larry Mason. "We go to work hard every day, twelve to fourteen hours a day, often six or seven days a week, raising our children, having Little League with our kids, thinking that we're apart from the world and very successful in our lives. Bam. The picture changes overnight. All of a sudden somebody's talking about a spotted owl."

Mason closed his family sawmill in Forks in 1990, one year before Federal Judge William Dwyer issued a temporary injunction barring timber sales in the Olympic National Forest until the Forest Service could come up with its belated spotted-owl protection plan.

"We who thought we were serving this nation, oftentimes risking our lives to do it, are told that we're forest rapers," says Mason, who now works as a spokesman for Alliance for America, a coalition that lobbies against environmental legislation. "Now this is not just an injustice because we've been robbed of our income, this is a false accusation. It's not dissimilar to the way America treated its veterans at the end of the Vietnam war."

The inability of the Forest Service to come up with a timely spotted-owl protection plan—and hence the judge's injunction—left the timber community of Forks, buried deep within the Olympic Peninsula, out in the cold. Unemployment among the 3,500 residents is soaring, families are stressed, and the people are angry. Social Services reports a 30 per cent increase in its caseload last year; domestic violence is up from one or two cases per month to seven or eight; mental-health workers are in demand, and the community food and clothing banks are being emptied.

In a round-table discussion on timber issues, members of the town shared their views. "This all hit us at one time," says Kaye Kelso, who grew up in Forks and logged for thirty years before retiring. "There was a day when you had to get a pair of track shoes to get across the highway here for all the trucks going across town. I mean you had to be damn fast to make it."

This January, the Forest Service released its spotted-owl protection plan, almost one-and-a-half years after a Congressionally imposed deadline. The plan, which sets aside 4.5 million acres of Northwest timer, has generated complaints from both industry and environmentalists. Estimates on how many jobs will be lost because of the plan range from some timer-industry projections of 180,000 to claims by some environmental groups that preserving the forest will result in a net gain of jobs, primarily in the recreation industry. The Forest Service estimates that 20,000 timber-industry workers in Washington, Oregon, and California will be put out of work under the plan.

But towns like forks will probably never see the booming economy of the 1970s and the late 1980s again. Increased mechanization and the processing of logs overseas account for much of the decline in the timber industry.

The old growth timber can be kept off-limits by protection or it can be logged off in the next decade, but the economic situation in Forks will remain bleak. Like most forestry statistics, estimates on the amount of old growth left on the Peninsula vary. The Wilderness Society performed a comprehensive study using Forest Service aerial photographs and satellite images, and concluded that all the remaining unprotected old growth on the Peninsula will be logged within ten to fifteen years at pre-1990 harvest rates. If these figures are correct, workers in the timber industry would be facing hard times soon—even if the spotted-owl controversy had never erupted.

Most of the Forks townspeople who participated in the round-table discussion seemed to think that all unprotected old growth should be cut. Some, like Kaye Kelso, subscribe to an old timber-industry view that "decadent" timber should be replaced by younger, faster-growing trees.

Mason echos Kelso's view. "There's a dynamic in nature," he says. "Man's gonna harvest that timber, or nature's gonna harvest that timber. If man doesn't cut it down, nature's gonna burn it down, blow it down, flood it down, one way or another. Or the trees are just going to die and fall over. Things are constantly in transition."

By cutting old growth and planting new trees, loggers identify themselves as responsible stewards of the land. "We're all environmentalists," says Glen Demorest, a retired logger, "because if we weren't, we couldn't enjoy what we're doing because it's very hard work."

Forks natives are as much victims of poor land management as the forest which provides their livelihood. In the mid-1980s, say residents, the Forest Service told the owners of small operations that the old-growth timber supply would last thirty years. Many owners invested in expensive and now-idle mill equipment tooled for old growth. In 1986, the Forest Service predicted a maximum sustained harvest of 185 million board-feet of timber. Today, that figure has been reduced to twelve to fifteen million.

The Forest Service failed to anticipate both the environmental controversy and one of its main causes, the massive liquidation of private and state timberland in the region. Throughout the 1980s, corporate landholders liquidated their timber supply, motivated by the climate of leveraged buy-outs, the easing of incentives to practice sustained forestry, the fear of future restrictions, and a lucrative market.

As a result, around Forks the environment is in peril and the pressure on the National Forests to accommodate logging, recreation, and wildlife is intense. Since the small companies rely primarily on Forest Service land for contracts, it is the local mill owner and the family-owned logging operation that have had to sell out.

"I don't think you can imagine what that does to your heart," says Mason. "You come home every day, you can't sleep at night, the bank is breathing down your neck, and for what? Because some politician several thousand miles away made a decision. And the decision was that he thought he would get more votes if he closed down the commercial forest than if he kept it open. Not because he thought it was immoral to cut trees, not because he had a better plan to manage the planet, but because he thought it would strategically be better for his re-election campaign."

Mason's new job at Alliance for America exemplifies the unlikely partnerships that are forming. A pro-timber coalition of corporate and union groups, it is a new and powerful lobby in Washington, D.C. Such labor members as the International Woodworkers of America are fully aware that the industry is simultaneously pressing Congress to allow permanent striker replacements. But their alignment with an industry with which it has engaged in bitter disputes over fair wages and benefits testifies to the frustration most timber workers feel.

Environmental groups are also forming tenuous alliances with workers to promote sustained-yield forestry and timber-community relief. In Forks, however, bumper stickers proclaim, LOGGERS: AN ENDANGERED SPECIES, and "preservationists" are not regarded kindly. Kaye Kelso says, "There's no way you can educate a fellow like that. He's biased and opinionated. You can't tell him anything. And he's got the whole country talking like that. I think environmentalists are going to shut so much down in this country that people are going to see them for what they are. That they are going to have to change some of these laws that pertain to endangered species and stuff like that to protect people instead of owls and rats and whatever they're protecting."

Not all loggers see environmentalists as a threat. Vince Beedle, twenty-eight, is a logger in Cle Elum, Washington. Unlike Forks, Cle Elum is located near a major highway and is supported by other industries. Beedle has held down other jobs, such as trucking, but he likes to log because of the good pay, the companionship in a small operation, and the outdoors. He and his twin brother are the youngest children in one of the largest families in Cle Elum. Beedle's father raised fourteen children as the head of a "gyppo" logging enterprise, a small operation that contracts out to larger companies.

"I think environmentalists have done a log of good in bringing the issues to light," Beedle says.

Most of the timberland around Cle Elum is second-growth white pine owned by private companies. Beedle showed examples of good and poor logging practices in the broad valley. He is proud of the operations that selectively log larger trees. The technique, suited for the gentle slopes and smaller timber, leaves the younger trees free to grow larger for later harvesting, and preserves aesthetics.

But, he says, some companies have been wiping out the forest in a rush for short-term profits. This familiar scenario created an artificially inflated timber economy and sacrificed future jobs and timber supplies. Beedle is reluctant to name names, but it's not hard to pinpoint Plum Creek, a Burlington Northern spin-off, as the prime culprit. In June 1990, *The Wall Street Journal* published a front-page analysis of Plum Creek"s intent to boost shareholder values by cashing out at the expense of the forest and the long-term interests of the people who log it. Only recently, now that most of its land is logged, said *The Journal*, has Plum Creek returned to responsible forestry.

The Endangered Species Act comes up for renewal this year, and there are many environmental measures working their way through Congress now. The Administration's handling of the spotted-owl issue points to what some, like Representative Peter DeFazio, Democrat of Oregon, call "crisis by design."

Administration policies have the effect of polarizing the environmental community against timber interests, thus heightening anti-environmental sentiment. In 1989, when the owl received a "threatened" listing, the Forest Service said it would develop an interim plan within thirty days to protect the bird. The plan never materialized. Congress then imposed a deadline of September 30, 1990, for a spotted-owl recovery plan. In late February 1992, Interior Secretary Manuel Lujan urged Congress to drop a key requirement of the Endangered Species Act. Instead of carrying out its legal obligation to do everything it can to increase the population of spotted owls, the Administration wants to do only what is necessary to maintain the owl population at its current level—which is near extinction. Lujan's plan will okay the logging of vast stands of ancient forest.

Last year, the Administration suppressed a report outlining options that could aid timber-dependent communities. Obtained anonymously by the Association of Forest Service Employees for Environmental Ethics, the report outlined options which included instituting retraining and economic assistance, banning all log exports (creating 15,000 new jobs), changing Forest Service contracting and purchasing authorities to help displaced workers, creating 20 per cent more wood supply by increasing harvesting and milling efficiency, and developing public-works projects including construction of recreation facilities, enhancing fisheries, gathering ecological data, reclaiming roads in owl-habitat areas, and forest monitoring.

Representative DeFazio told reporters that the report was suppressed by "someone in the White House."

But families in Forks have been reluctant to take even the small measure of government-sponsored assistance now available. "I think you'll find one thing very unique about the people who work in the timber industry," says Jack Zaccardo, a forester employed by the state of Washington. "They are not people to sit around on their duff. They work hard, they earn an honest living, they do not want to take anything like welfare or unemployment if they can help it."

Many residents remain optimistic that good times will return to Forks. But one man in Forks offers what is probably a more realistic outlook. "Unfortunately," says Glen Demorest, "I see it as the entire country has come to an end of an era of people who work for a living. And when you shut down the loggers, the coal miners, the farms, the oil well people, and everyone else, we're not going to have many producers left. Solutions need to come quickly, and I don't think they will come quickly enough for a lot of people."

— DAVE FISHER AND
CHARLOTTE SCHUBERT

(Dave Fisher, a schoolteacher in Seattle, conducted the interviews in Forks for the radio program "Earth on the Air.")

Discussion for Endangered in the Northwest

1. The Northern Spotted Owl issue has forced people to examine the Endangered Species Act which currently does not allow economic factors to influence whether or not a species deserves protection. Do you feel economics should be considered when determining the fate of a species? Give reasons for you views.

2. Many people feel that saving the owl is not really the issue, but the entire forest ecosystem is actually what must be saved. In your opinion, should the Endangered Species Act be changed to save ecosystems, not individual species?

3. "Timber Towns in Trouble" describes the loggers as victims. The environmental article makes the owl out to be the victim! Who is the true victim?

4. Before the captive breeding program was initiated, scientists studying the small wild population of the black footed ferrets took several measures to reduce the risk of spreading disease to the animals. How did they accomplish this? Do you feel these efforts were too extreme?

5. Why is it important to have more than one population of an endangered animal?

6. Many biologists claim amphibia are the 'canaries in the global coal mine.' What do they mean? Do you think there is enough evidence to support this claim?

7. In the Rocky Mountains populations of wood frogs are increasing and the northern leopard frog is decreasing. What could be the cause? Does this disprove the 'canary' theory?

8. Explain how acid rain and global warming can adversely affect survival of amphibians. Why are amphibians so vulnerable?

9. If amphibians disappear what will be the effect on the ecosystem?

10. Make a list of those animals that should, in your opinion, be saved. Prepare a defense for each one on your list.

Introduction

Marine Mammals... Cute But With a Price

This exercise investigates the problems with two of our well known marine mammals, the sea otter and the sea lion. One is cuddly with a public image that is hard to beat and the other relies more upon the scientific data for its protection rationale. Both are fully protected but the conflict between them and people are putting strain upon the protected species laws.The sea otter is not just a small adorable animal, it is a feeding machine that consumes greater than 20% of its body weight in food each day. Favorite foods include bivalves, sea urchins, abalone, and crustaceans. The sea otter (Enhydra lutris), once had a range from the tip of Baja California to the Aleutian Islands in Alaska. Unfortunately, their pelt has been such a valued commodity that they were almost trapped into extinction by American and Russian fur traders. By 1914 the total otter population was estimated to be approximately 100 animals. Action was taken in 1911 in the form of an international treaty prohibiting the trade of otter pelts and in 1973 the effort was further supported by the U.S. Marine Protection Act. Today, the otter population in Alaska ranges from 100,000 to 200,000 animals and 1,600 animals in California. It is evident that the populations are recovering very well. They continue to slowly push their way south invading the territory they once occupied. With their expansion they bring several concerns to government, special interest groups, and private industry.

AN ECOLOGICAL OVERVIEW

The otter is a typical "keystone predator." It feeds intensely on certain foods ultimately impacting the whole community. In the northern Pacific waters the otters are wiping out populations of invertebrates particularly those of the sea urchins (*Strongylocentrotus spp.*). The sea urchins are herbivores that feed primarily on settling single celled algae, diatoms and especially large kelp. Kelp forests house a very complex community of sea life. Kelp slows wave action, ocean current and provides an abundance of shelter. If this type of community expanded what are some accompanying ecological implications?

George Wirtes

James Wilson

ECONOMICAL CONCERNS

The fishing industry has long been upset with the sea otters and their effect on the market. The otters' appetites are causing populations of popular fish communities to dwindle and this is resulting in less profits. Japan has proven to be a strong market for Sea Urchin roe. Japanese New Year prices were about $2.30 per pound and some divers made more than $5,000 for a days work! Regular mid year prices are about 60 cents -$1.00 per pound.

NO OTTER ZONATION

The Pacific Coast Federation of Fishermen's Associations (PCFFA) has seven point plan to minimize the sea otter problem. It is designed to establish boundaries where otters are prohibited. The proposal states that any otter found wandering beyond the border will be captured (by non lethal means) and relocated to another suitable area. There are several beneficial points to this proposal such as:

1) Keeping the otter population away from high oil vessel traffic thus minimizing the chance of an oil spill coming in contact with them and limiting oil leases in the area.
2) Establishing an "Otter Monitoring and Habitat Evaluation Center."
3) Establishing shellfish hatcheries in the otter zones.
4) Banning all dumping in otter habitat.

There are also several obvious points that need to be taken in consideration. With the otter population constantly expanding, are we able to hold them back keeping the area in fact "otter free"? Moral issues cannot be avoided in this case. Are we acting in the best interest of the environment? Can we set limits on something as wild as a marine mammal?

GOVERNMENT ACTION

As a solution to ecological and economical pressures the government has expended $300,000 in an effort to relocate otters to San Nicholas Island. Unfortunately, the plan has not worked so well. Many have made the intense journey back to their home land and several have died. The United States Fish and Wildlife Service are disappointed with the results but hope the plan will soon take root and the population will soon flourish. Is this the most favorable action to take? Relocating a species is a big gamble. There are so many variables to take into consideration. One is the new species effect on adjacent communities. Will they compete with other marine life for the same foods in effect driving a species from the area? This is entirely possible given that starfish and otters have similar feeding patterns. Another important variable to take into consideration is the inhabitability of the new environment. Will the introduced otter feel comfortable enough with the area to stay and continue flourish? Relocation is more than transferring a species and hoping that it will adapt. Relocation is when it is asked to develop or redefine a niche.

THE SEA LION

In 1972 the sea lion was placed on the endangered species list. This is the toughest legislation the government has to protect species. Prior to this the sea lion was considered fair game and the fishermen and in some instances even the Marine fisheries departments killed sea lions to protect the fishing resources. In British Columbia a massacre of sea lions was undertaken with a machine gun mounted upon the bow of a fisheries patrol boat. This control together with the commercial harvesting of the seal put these marine mammals in danger of extinction.

MERCHANTS vs. MOTHER NATURE

In San Francisco, California, a year ago, sea lions were occasionally sitting on docks and basking in the sun. Because of the warm weather and abundance of food, the sea lions adopted dock K, pier 39 as their own. They frolic and feed the entire day and leave temporarily each season to mate. Tourists now come to feed about 500 animals that laze on the docks. Many merchants and yacht owners however, do not welcome the animals. They maintain that the animals smell very foul and bark continuously throughout the night. The sea lions, however, are protected by 1972 legislation.

When hordes of sea lions found easy pickings at the Ballard Ship Locks during the annual steelhead run officials attempted to remedy the situation. Six sea lions were captured and trucked to Santa Barbara, California, and then by boat to San Miguel island. Within months most of these transfers were back! This exercise cost $15,000.

GOVERNMENT ACTIONS

The National Marine Fisheries has devised a plan that permits some killing of Marine mammals for certain class of fishing. Under the draft plan, if the fishing industry is adversely impacted by marine mammals then a certain number of the marine mammals can be culled. This number will include those removed by researchers, aquariums and zoos.

Safe havens for our marine mammals are becoming harder and harder to find. Recreation, fishing and industry are encroaching on our wildlife habitat. It is nice to see action that serves to protect the vanishing habitat and the Puget Sound venture has to be one of the biggest.

Questions and Concerns

READING 1

When reading this article consider the following:

The population is slowly growing without too much help from the government so do we really have to 'protect' these animals?

Reading 1

Reintroducing a Political Animal

What happens when the government moves sea otters around in California? Nobody is happy, including the otters, which try to swim home. Many attempts to relocate threatened species fail.

Big Sur, California

TO PROTECT THE SEA OTTER from the ravages of man, government biologists are committed to scooping as many as 250 otters from their home waters off the jagged coast of Big Sur and flying them to an isolated island southwest of Los Angeles, where it is hoped the transplanted animals will establish a new and happy home. Unfortunately, the sea otters have not read the script.

Of the 63 otters carted to San Nicolas Island during the past 11 months, less than a fourth remain. The older and stronger animals swam back to the mainland. About half of the relocated otters have vanished. The rest went to Davy Jones's locker.

It seems that saving a threatened species can be a controversial and disheartening business, fraught with political peril and scientific strife. "Nobody realizes how frustrating these kinds of things can be....Most reintroductions that are attempted, fail, and we never know why," says Galen Rathbun, a biologist with the U.S. Fish and Wildlife Service.

Despite such setbacks, the practice of reintroducing a species into a remnant of its former range has become a tool of choice among conservationists and wildlife managers. According to an unpublished survey by Michael Scott and Brad Griffith of the U.S. Fish and Wildlife Service and the University of Idaho, there are at least 500 reintroduction episodes each year in North America. Though the majority of these are aimed at establishing populations of game animals such as big horn sheep and wild turkeys (populations which are later culled by hunters), there are active or planned reintroductions for at least some of the 495 species listed as threatened or endangered in the United States. Particular attention is being paid to such glamorous subjects as bald eagles, red wolves, and sea otters. The U.S. Fish and Wildlife Service will spend about $30 million this year on activities related to species in peril.

No one wants to be against something as hopeful as the reintroduction of wildlife. For what could be wrong with alleviating some of the damage brought about by habitat destruction and exploitation? Thanks to reintroduction programs, for example, there are now 20 pairs of bald eagles nesting in Texas and eight red wolves loping through a swamp in North Carolina. But whether these costly and continuing reintroductions will succeed is unknown.

Biologists confess there are a number of problems with the practice. First of all, reintroductions often do not work. According to Scott and Griffith, of those reintroductions that can be judged, less than half succeed. For instance, the wildlife service has spent over $1.5 million since 1975 in a futile attempt to create a self-sustaining population of whooping cranes at Gray's Lake in Idaho. This spring, there were only 16 whooping cranes in the flock. None of the cranes have produced offspring, despite 13 years of encouragement, reports Jim Lewis, project coordinator for the wildlife service.

Says Tom Cade of the Peregrine Fund in Boise, Idaho: "Reintroductions are last-ditch and desperate efforts.... They can work, but they take lots of time and lots of money."

Certainly, a few reintroductions are working. Cade's falcons are a case in point. Nearly 2500 peregrines have been released into the wild since 1975. On the East Coast, there are now between 65 and 75 territories populated by captive-reared and released falcons. Says Cade: "In a few more years, we'll be back to where the population was before DDT took them down."

Yet, a continuing problem with reintroductions is that biologists must often contend with manipulating a dwindling species they do not fully understand. Wild animals in wild settings have a way of upsetting the best laid plans. Even though a great deal is known about the natural history of the peregrine falcon, for example, who could have guessed that falcons reintroduced into the wild would actually prefer eating pigeons and living in cities like Baltimore? As for the otters, the wildlife service never believed that so many otters would leave San Nicolas Island.

By all accounts, little has gone according to the elaborate Recovery Plan written for the sea otter. For starters, catching an otter in a dip net is not like plucking dead guppies out of an aquarium. Government biologists recently discovered that if a sea otter has been caught once, it is extremely difficult to catch again, particularly when the otter has a government-issue radio transmitter surgically implanted in its belly. Some otters have become so wary that the biologists have resorted to dispatching teams of Navy-trained scuba divers outfitted with special bubbleless rebreathers to sneak up on the otters while they are asleep. Even so, of the 30 otters sporting a surgical implant, only three could be recaptured and sent to San Nicolas. Once out at the island, all three eventually disappeared, thereby completely foiling a major experiment designed to compare the life-styles of mainland otters and transplanted otters.

In all, about half of the otters have vanished from San Nicolas Island. Upon release, one confused animal immediately began swimming in the direction of Hawaii. He has not been seen since. Most of the older otters also left. The lucky ones made it back to their breeding territories along the mainland, a remarkable swim across 200 miles, and a journey that is teaching the biologists

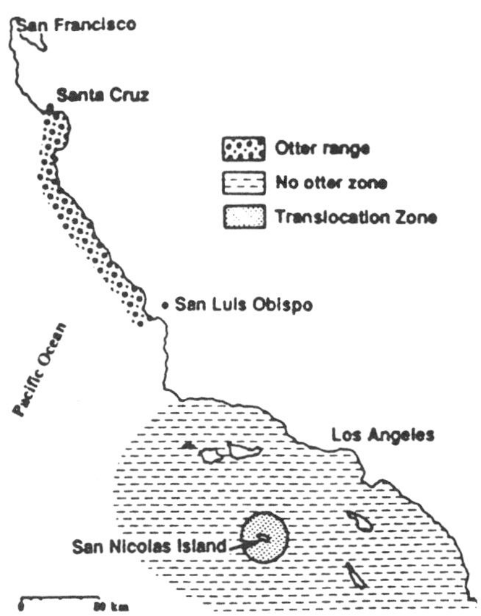

Holly Bishop

something about the strong fidelity felt by otters for a particular home range.

At least ten otters are confirmed dead, the victims of stress or drownings or foul play. A few months ago, one of the relocated otters washed up near Point Mugu in southern California. She had a bullet in her head and chains wrapped around her body. Says Rathbun: "It looked like a gangland slaying."

Killing sea otters is nothing new, but the death of one of the reintroduced animals illustrates how emotional such a program can be. Like the gray wolf and the bald eagle, the sea otter has come to represent far more than just a unique set of genes. "The sea otter is a symbol," says Robert Brownell, a fish and wildlife service biologist heading the otter reintroduction program.

To the fishermen of California, the otters are seen as direct competitors that consume such commercial delicacies as pink abalone, red sea urchins, and spiny lobsters. To the oil industry, which hopes to drill wells off the coast of central California someday, the otters represent a potential stumbling block, since their federally protected status could hamper development. To the animals' advocates, otters are dewy-eyed and cuddly creatures that must be defended at all costs. And to the government biologists charged with protecting and managing the species, sea otters are proving to be fascinating and exasperating wards of the state. Says James Estes, a fish and wildlife service biologist at the University of California at Santa Cruz: "We were never trained to work in an arena where everything is controversial and everything is political."

The range of the sea otter, *Enhydra lutris,* was once the entire northern arc of the Pacific Rim, from Hokkaido in northern Japan, through the Kuril Islands and the Kamchatka Peninsula in the Soviet Union, across the Aleutian Island of Alaska, and down the west coast of North America to the middle of the Baja Peninsula. But unfortunately, the otter's dense and supple pelt was much admired, particularly by the Chinese, who had an almost maniacal craving for the otter's fur, which they used as trim on their coats. A log entry from the Russian fur ship *Ilmen* notes that in 1815 Captain Boris Tarasov and his crew obtained 955 otter skins during the 7 months they were stationed at San Nicolas Island. The bounty did not last for long. A diary entry from a Santa Barbara fur hunter noted that there were so few otters out at San Nicolas Island in the 1850s that it was hardly worth the trip.

By the turn of the century, the sea otter was so greatly exploited that some thought the species was extinct. By 1914, the California population is estimated to have been reduced to no more than 100 otters.

Since then, the otters have made a remarkable recovery. Today, there are more than 100,000 otters living in the Aleutian Islands of Alaska and 1,650 living in California, where the animal's range has slowly spread from Big Sur to include 200 miles of coastline from San Luis Obispo in the south to Santa Cruz in the north.

Yet despite such gains, the wildlife service considers the California sea otter population in a precarious position because more than 100 million barrels of oil travel through the sea otter's range each year. To protect the otters, the service called for the establishment of a second population of animals as a way of hedging the bet against devastation from a single spill.

Located about 70 miles southwest of Los Angeles, San Nicolas Island belongs to the Navy's Point Mugu Pacific Missile Test Area. Though the island is off-limits to civilians, its rich waters are actively fished. Pinning down the dollar value for the lost revenues due to the relocation of otters is a tricky thing. Gross annual loss of the shell-fisheries is estimated to be between $106,000 and $264,000, according to the California Fish and Game Department. But Steve Rebuck, a spokesman for the group Save Our Shellfish, calls these figures "a joke....The numbers are ten times that amount." Rebuck represents the anger and distrust felt by commercial fishermen in California when he refers to the entire otter project as "a hoax."

This ill will has led to a series of compromises between conservationists and the fishing industry. It was agreed that the otters could have San Nicolas Island but the rest of southern California south of Point Conception would be maintained as a "No Otter Zone." Any otter found in southern California is to be captured by a kind of otter-SWAT team and returned either to San Nicolas Island or the mainland population.

"It must be said: this is not a conservationist's dream. It's a compromise and like all compromises there are things we don't like about it," says Rachel Saunders, staff biologist for Friends of the Otters, a powerful 4500-member organization which follows with keen interest anything to do with threatened otters.

Neither are the fishermen happy. "The No Otter Zone is a bunch of baloney," says Rebuck, who contends that the mainland of southern California would never support many otters anyway because of the area's wide sandy beaches and lack of otter chow.

It may not matter. In the long run, the No Otter Zone may prove a difficult artifice to maintain. It is one thing to snatch a few truant otters that stray from San Nicolas Island. It is another to stop the mainland population's inexorable push south. Already a large group of males is within 70 miles of the No Otter Zone at Point Conception.

"All indications are that the mainland otters will continue to disperse and if you have to contain them, you're going to have to cull them. But to even suggest culling otters in California right now would be insane," says Estes. Instead, there is talk of sterilization and of implanting a birth control device in female otters similar to the ones used for deer. Also, the California Department of Fish and Game is experimenting with taking otters from the southern portion of their range and moving them to the northern edge, a kind of shell game that cannot last forever.

As for the success of the otter reintroduction at San Nicolas Island, it is far too early to tell. "If I was a betting man, I'm not sure which way I'd go," says Estes. The government is betting about $300,000 a year on the effort. Certainly, the biologists were surprised and dismayed to watch the island population dwindle. But the otters that remain seem to be happy, says Rathbun. It often takes years before a population is self-sustaining.

Starting next month, the government biologists will begin to capture as many as 70 otters to add to the small group left at San Nicolas. This time they will take mostly juveniles, which appear less likely to leave.

At the very least, the sea otter has taught biologists something about the difficulty of manipulating nature. Says Rathbun: "Sometimes, in my more private moments, I wonder what we're really doing for the critters."

Some scientists are asking a similar question on a more general level by challenging the wisdom of allowing conservation efforts to be driven by a single endangered species rather than an entire ecosystem.

"We've become so focused on preserving a species here and a species there that we are losing sight of the habitat," says Richard Hutto of the University of Montana in Missoula. Hutto believes this narrow focus by the government on "management indicator species" is obscuring the real importance of preserving what Hutto calls "the integrity of patterns and processes" of an ecosystem. Says Hutto: "The species-dominated approach gives us the illusion that we can keep growing and keep destroying and then just mitigate the damage later with a reintroduction....It's arrogance."

Cade of the Peregrine Fund thinks that Hutto may be missing the point. "When you save a predator, everything that the predator needs to survive, goes along with it," says Cade. This approach has been called "trickle-down conservation" by some researchers.

Many conservationists would agree with Hutto in part, however. "Saving endangered or indicator species is not the same as preserving biological diversity," notes Scott of the wildlife service and his colleagues in a reply to Hutto's critique in the pages of Endangered Species Update. Pragmatists though, Scott and his fellow authors point out that "it is easier to argue for funding to save a charismatic species than to save the abstractions we call ecosystems. It is no accident that there are only 13 insects among the 967 taxa listed in 1988 as threatened or endangered by the U.S. Fish and Wildlife Service."

■ WILLIAM BOOTH

The Otter-Urchin-Kelp Scenario

The reintroduction of sea otters to an isolated island off the coast of Los Angeles may provide researchers with a tidy laboratory in which to peel apart the complex and much debated interplay of otters, urchins, and kelp forests.

Since 1980, marine ecologists have been monitoring intertidal pools and subtidal habitats at San Nicolas Island by screwing stainless steel eyebolts into the sea floor and returning periodically to the same swaths of rocky benthos to record whatever happens to be there, be it encrusting algae, finfish, urchins, or kelp. Now for the first time in recent history there are also some otters to look at.

It is well known that sea urchins often wander about—albeit very slowly—to graze upon the holdfasts of kelp. Heavy overgrazing upon the macroalgae by the herbivorous echinoids have created denuded "urchin barrens." Think of urchins as a kind of goat of the sea.

It is also understood that sea otters have voracious appetites and are capable of ingesting about one-fourth of their body weight a day. Otters, like sushi gourmands, have a great fondness for the gonads of sea urchins. In the economies of foraging, urchins provide a valued and accessible prey item, since the otters need only dive to the bottom and pick them up—unlike an abalone, for example, which often requires repeated dives and use of a big rock to beat upon the creature until it loosens its hold.

The question, then, is do sea otters maintain the kelp forests by consuming the urchins that consume the kelp?*

James Estes and Glen VanBlaricom of the U.S. Fish and Wildlife Service in Santa Cruz suspect that otters have a profound and direct impact on kelp forests. Estes and his colleagues find support for the hypothesis by comparing areas with and without sea otters, by examining the historical records, and by mimicking otter predation with experimental patches of sea floor where all the urchins have been removed by man.

Estes says that in southeastern Alaska, for example, a population of otters moving into a new area has an immediate impact on urchins, followed by a rapid recovery of kelp. In cases where the urchins have been removed by researchers, Estes and Christopher Harrold of the Monterey Bay Aquarium report that "community structure in the experimental plots shifted from urchin barrens to algae dominated assemblages, whereas unmanipulated control plots remained unchanged."

But Michael Foster of Moss Landing Marine Laboratories in Moss Landing, California, disagrees with Estes and argues that the sea otter is not a "keystone" predator, but "another brick in the wall" in a highly complex system that is often in flux. Foster and David Schiel of the Fisheries Research Centre in New Zealand reviewed data from 220 sites in California outside the otter's range and found that deforestation by sea urchins is the exception rather than the rule. Dense forests of kelp often exist side by side with urchin barrens. Why? Foster believes that the otter-urchin-kelp hypothesis is overly simplistic and that other factors such as water motion, light, nutrient levels, and substratum are as important as sea otters.

The story has another twist. Urchins have two ways to make their living: they can actively graze or they can sit in crevices and wait for pieces of kelp to drift by. Harrold and Daniel Reed of the University of California at Santa Cruz found that "the intensity of grazing was independent of sea urchin density." What counted most was the amount of kelp that drifted by. In *Ecology* the two report: "At the beginning of the study the biomass of drift algae in the barren site was not sufficient to keep urchins well fed (as indicated by their low gonad indices). As a result, they actively grazed the substratum, preventing the establishment of macroalgae....Once drift algae became available to the urchins, they switched from an active, grazing mode of feeding to a sedentary drift-feeding mode."

Harvesting kelp is $50-million-a-year business in California, while kelp forests also support a rich community of marine life, including a number of commercially exploited fish. To some extent, conservationists have sold the otter to resource managers and fishermen as an animal that could increase the vigor of the kelp forests, thereby encouraging a whole cascade of events. Out at San Nicolas Island, firmer answers may be forthcoming.

■ W.B.

***The Community Ecology of Sea Otters,* edited by G. R. VanBlaricom and J.A. Estes (Springer-Verlag, Berlin and New York, 1988).

Reading 2

Questions and Concerns

READING 2

When reading this article consider the following:

1. To whom are these creatures adorable?

2. In what ways are they like the bears of Yosemite?

ENVIRONMENT

Sitting on the Dock of the Bay

The boys of blubber return to a San Francisco pier

They stink, they snore, they slobber. And they're adorable, San Francisco's noisiest waterfront attraction: 150 blubbery, belching male sea lions who've set up housekeeping on K-Dock near Fisherman's Wharf.

Last January, an armada of the 800-pound mammals invaded the boat slips of Pier 39. And why not? The fishing was great (more herring than even they could eat), the marina was calm, and oh-those-California-rays at high noon! The good news traveled fast: One day, there were six sea lions; the next morning, 100, and four times that within a month. Tourists came in waves and business at Pier 39 — a collection of souvenir shops and fooderies — picked up. By spring, though, the animals departed — presumably to find their sea lionesses and new pups on the Channel Islands to the south.

Now the whiskered fellows are back, hauled up on the K-Dock as if they never were away. "As far as the sea lions are concerned," says Fritz Arko, president of Pier 39, "this is home." Arko can barely make himself heard, as he proudly points to the pungent pinnipeds carrying on just 50 feet below beyond a locked gate. They're constantly barking and yelping as they shove one another off the dock in an aquatic version of King of the Hill. The clever ones try sneak attacks by swimming under the dock to blindside the bigger players; they usually get pummeled. "It's more exciting than an aquarium," says 8-year-old Neal Burgner, visiting with his mother.

Not everybody is amused with the antics — and hygienic habits — of the returning residents. Their nocturnal barking, which can be heard miles away, infuriates nearby yacht owners. Their droppings are the size of grapefruits, and the lingering smell makes herring seem positively perfumy. And perhaps most disgusting of all is the sea-lion expectoration. Pier management is trying to help. Workers now regularly scrub and deodorize the play areas. Additional pier space has been built. But even if most locals agreed it was time for the lions to leave, there is little that could be done. The marine animals are protected by a 1972 federal law, a statute as tough as any tenants' rights provision, which bars the harassment of these seaworthy squatters.

Experts worry that the sea lions will become like the bears of Yosemite: too used to people and too lazy to hunt for food. "They're slugs," says Brian Gibeson, a Bay Area marine biologist. "They'd do anything for fish and Pier 39 is the ultimate deli." But whether they stick around or return to their natural migratory habits along the Pacific coast, the sea lions have been immortalized in a truly San Franciscan way. Intrigued by their sounds, composer Doug McKechnie recorded the best barks of the bay and used computer technology to produce a song that would make Tony Bennett envious. Soon playing at a wharf near you: "I Left My Arf in San Francisco."

DAVID A. KAPLAN *with* NADINE JOSEPH *on Pier 39*

Questions and Concerns

READING 3

When reading this article consider the following:

What is happening to these creatures and who do you think is doing it?

Seizure Activity in Sea Lions

by Dr. Judie Gerber

Ten California sea lions *(Zalophus californianus)* with seizure activity have been admitted to The Marine Mammal Center since this study began in January 1991.

The first sea lion admitted was a 97-pound yearling female which stranded in January 1991 near Santa Cruz. This animal had seizure activity throughout her hospitalization period. She was afraid of water, partially blind, and displayed abnormal behaviors when interacting with pen mates. Unfortunately, she did not respond to treatment and her health deteriorated rapidly. She was euthanatized in April. On gross post-mortem examination, the only obvious lesion was the pale, tan, friable liver with pinpoint white foci randomly distributed on the external surface. Toxin testing of liver tissue revealed elevated mercury levels. Mercury in the liver likely reflects exposure to elevated mercury in the environment or from ingesting contaminated fish. The effects of long-term, low-level mercury exposure in marine mammals is not completely understood. This sea lion's brain, liver and kidney tissue also contained slightly elevated levels of DDE (a metabolite of the pesticide DDT), indicating environmental exposure.

Weighing in at 276 pounds, a subadult male stranded in February in Santa Cruz. He had seizure activity until the end of July and was released in mid-August. Diagnostic testing revealed a whole blood selenium level of 2.054 parts per million (ppm). The expected selenium range in mammals is 0.08 to 1.0 ppm. Selenium, a toxic trace element known to cause seizures and other ailments, occurs naturally in California in certain soils, and in heavy crude oils found in the San Joaquin Valley and in portions of offshore oil deposits. The two major sources of selenium in Northern California are discharges from oil refining and unidentified discharges through public sewers. Elevated whole blood selenium levels in the body indicate excessive exposure to selenium in food or the environment.

Another subadult male stranded in June in Monterey. This 81-pound animal had a seizure upon admission to The Center, but did not seizure after that initial episode. He was released in August. Again, the blood selenium level was slightly elevated at 1.577 ppm.

September was the big month for seizure animals. On September 18, three female adult California sea lions stranded in San Luis Obispo. One was dead on arrival. Gross post-mortem examination revealed no obvious abnormalities; however, electron microscopy of the choroid plexus (vascular red tufted tissue which produces cerebral spinal fluid) of the brain, revealed the presence of a viral particles, more specifically, a rotavirus. The significance of the virus is not known. Post mortem tissues contained increased selenium levels at nontoxic concentrations for most mammals. The other two female sea lions were luckier. One had seizures for two weeks, and the other for three

Courtesy of "The Release," the quarterly newsletter of The Marine Mammal Center, Sausalito, CA weeks. Both animals had very high whole blood selenium levels (One had 5.99 ppm and the other had 2.97 ppm.) upon admission. With supportive medical care for the first few weeks of hospitalization, the blood selenium levels dropped and seizure activity ceased. Both animals were released in November.

On September 28, three more animals were rescued by The Center because they had seizure activity on the beach. All three animals were dead on arrival. An 82-pound subadult stranded in Santa Cruz; a 110-pound male subadult stranded in Monterey; and the third DOA, a 209-pound male adult, stranded in Humboldt. Post mortem examination of all three sea lions revealed no significant lesions, however, the Monterey animal's histopathology results showed a severe nephritis suggestive of leptospirosis. Animals with kidney failure will display seizure activity in the last stages of the disease. The liver in all three animals had nontoxic background levels of DDT and its metabolite, DDE. Post-mortem tissues also contained elevated selenium concentrations.

Stomach contents and samples of the large and small intestines from the animals rescued from Monterey and Santa Cruz were sent to the California Department of Fish and Game. In September 1991, a large number of brown pelicans had died in the Santa Cruz area. These birds exhibited signs of a central nervous system disorder before death. Post-mortem examination of the dead birds by Fish and Game revealed no significant lesions, and pesticide and heavy metal levels were not abnormally elevated. Samples tested for the possible presence of red tide toxins suggested that the neurotoxin, domoic acid, was causing the mortalities. Utilizing recently developed methods of analysis, domoic acid was found in the pelicans' stomach contents and anchovies (the principal food source at the time), and no domoic acid was found in normal, healthy pelicans. The stomach contents from the two sea lions were negative for domoic acid; however, we now routinely save stomach contents from seizure animals for future studies.

The investigation of seizure activity in California sea lions has provided The Center's medical staff with valuable information about seizure etiology in our pinniped patients. Suspected causes of seizure activity include man-made and marine toxins, and viral agents. Animals with severe diseases such as renal failure, also exhibit seizure activity. The Marine Mammal Center is committed to this ongoing study. We will continue to fully evaluate sea lions with seizures in order to expand our knowledge of the ocean environment and improve diagnostic and therapeutic regimens at The Center. ❖

Dr. Gerber is the 1991 Veterinary Fellow.

Reading 4

Questions and Concerns

READING 4

When reading this article consider the following:

Gunshots are not the only thing that impacts the sea lion. Consider the chemicals that are present in our waters and consider the effects upon the animals that live in that water.

Fair Game?

by Denize Springer

Early on the morning of July 6, 1992 gunshots broke the silence over Monterey Bay in the vicinity of Monterey, California's famous Cannery Row. Acting on information provided by The Marine Mammal Center, law enforcement officers from the National Marine Fisheries Service (NMFS) and the California Department of Fish and Game watched from their vessel. Between 2 and 4 a.m., the officials witnessed a local squid fisherman shooting at sea lions swimming near his boat. Once aboard the fishing vessel, the officers seized a .22 caliber semi-automatic rifle, 290 live rounds of ammunition, a number of spent shell casings and a 30-round magazine. Three and a half weeks later, the squid fisherman was charged with nine counts of violating the Marine Mammal Protection Act (MMPA).

Unprecedented Number of Animals Shot

As this year's El Niño conditions brought more than the usual number of sea lion yearlings perilously close to human activity, medical staff at The Center assumed they'd see a few more than the usual count of injuries from interaction with nets, propellers and guns. Mere assumptions, however, meant little when the number of shooting victims grew to unprecedented proportions. Over a period of the past eight years, The Center treated only 37 confirmed shooting victims. In the first eight months of 1992 alone, The Center admitted 75 maimed, blinded and permanently paralyzed sea lions bearing the evidence of everything from birdshot to large bullets. The majority of these animals died or had to be euthanatized.

Ninety-five percent of these animals were rescued along the central California coastline. All but three were California sea lions. Between mid-May and mid-June alone, 22 sea lions suffering gunshot injuries were brought in from Santa Cruz and Monterey counties. The Center's science staff worked day and night. In the stomachs of some of the dead were the remains of squid, a common prey item of many marine mammals. Lodged in brains and spines were .22 caliber bullets. The latter were turned over to forensic experts at NMFS.

In response, The Center reactivated its Marine Mammal Shooting Reward Fund and printed new posters. With pledges from associates and businesses, The Center was able to offer a $20,000 reward for anyone giving information that leads to the arrest and conviction of anyone caught "illegally" shooting marine mammals.

Protection Act Allows Killing

In some instances, it is "legal" to shoot marine mammals. Ironically, it is the MMPA that provides commercial fishermen with allowances to kill marine mammal species (other than threatened, endangered or depleted stocks) that are a serious threat to a vessel's equipment, catch or persons aboard. Such permits are issued to certain commercial fisheries for net fishing as well as some line fishing party boat operations. The regulation requires a sequence of actions however, that these fishermen must follow to deter marine mammals from damaging gear or catch. It emphasizes that all non-injurious methods available must be taken <u>before</u> intentionally injuring or killing a marine mammal. It also states that an animal must be in the process of damaging a fisherman's

Courtesy of "The Release," the quarterly newsletter of The Marine Mammal Center, Sausalito, CA gear or catch before any harassment action may be initiated. The certificate states:

The certificate does not authorize the initial use of deadly force. If this is the first means of deterrence used, it will be considered a violation of the regulations and could result in a fine or suspension of the Certificate.

Almost all of the sea lions affected and rescued by The Center were small, 25 to 40 pound yearlings that were having trouble finding food near the water's surface, warmed to an unusual temperature by El Niño. Dr. Kimberlee Beckmen, The Center's Veterinary Fellow who is in charge of necropsy and examined all the animals affected says, "How could anyone think that these animals — not bigger than the family dog — were much of a threat to anyone or anything?"

The owner of the aforementioned squid boat obviously felt differently and misinterpreted or ignored the MMPA's allowances for commercial fisheries. Based on the eyewitness accounts of the California Fish and Game and NMFS officers, the fisherman was charged

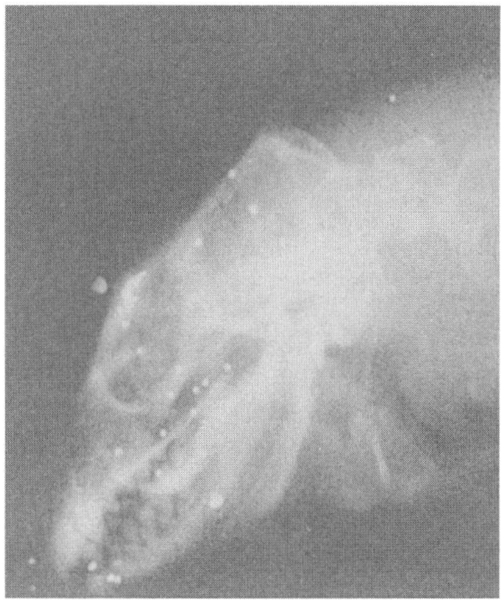

X-ray shows sea lion shot with three different types of ammunition

with three civil counts of shooting at marine mammals while none of his fishing gear was deployed, one civil count of shooting at marine mammals with no equipment deployed without first trying other means of deterrence, four civil counts of shooting at marine mammals while gear was deployed without first trying other methods of deterrence and one additional civil count for failure to report/log and notify NMFS of marine mammal shootings. The NOAA General Counsel assessed the fisherman's fines at $40,000.

The provision that allows commercial fisheries to shoot marine mammals was evidently offered as a compromise to fisheries people who felt that the MMPA went too far and in the process would eventually slow or curtail their business significantly. Sea lions, among other species are opportunistic feeders that eat many different kinds of fish and squid. As sea lions are a coastal species, known for their cleverness and learned behavior, such as following fishing boats, it is not hard to see why this species is the one most injured by guns and most openly hated by some humans who also compete for food from the sea. However, while some may view the currently healthy population (approximately 180,000 worldwide) of California sea lions as a detriment to the human demand for seafood, it cannot be denied that these animals are nothing less or more than skilled, intelligent competitors for resources in their natural environment ❖

Denize Springer, Communications Officer

Reading 5

Questions and Concerns

READING 5

When reading this article consider the following:

What is this area like? What are the other activities going on in this area?

Northern Puget Sound Sanctuary Taking Shape

The National Oceanic and Atmospheric Administration (NOAA) is making headway in its assessment of northern Puget Sound for possible designation as a national marine sanctuary. This remarkable site is known for its scenic beauty and productive estuarine waters, providing food and shelter for many species of fish, birds, and marine mammals. Northern Puget Sound also supports extensive shipping, fishing, and recreational activity.

If all proceeds according to schedule, the Draft Environmental Impact Statement and Management Plan for the site should be released for public review this summer. The coming months will be critical in shaping the scope and integrity of this sanctuary proposal, and in determining whether or not it goes forward.

Although Congress directed NOAA to prepare a prospectus for this site, it did not mandate sanctuary designation. Since the proposed sanctuary is located completely in state waters, Washington's Governor must also approve of the proposal. The State's support is thus key, as is backing from Washington's congressional delegation.

More than 1,000 people (including CMC staff) attended regional "scoping" meetings held by NOAA in November 1989. The following spring, NOAA and the Washington State Department of Ecology formed a "working/advisory committee" to help further define the marine resource issues of concern and what actions the proposed sanctuary could take.

Representatives from federal agencies, state, local and tribal governments, as well as user and public interest groups (including CMC) have been involved in the work of the advisory committee and its eight subcommittees (administration, education, research, water quality, habitat management, shipping and transportation, fishing and aquaculture, and tourism and recreation).

NOAA is reviewing a preliminary draft proposal, including administrative, boundary and regulatory alternatives. The agency has identifies six boundary alternatives thus far, ranging from several small, noncontiguous sanctuary zones surrounding selected islands to a sanctuary encompassing the entire study area.

Key management concerns include improving prevention, response and mitigation for such hazards as the transportation of oil, establishing an effective dialogue with Canada to address water quality and other "transboundary" issues, and minimizing impacts to wildlife from boating and air traffic.

The contents of NOAA's proposal are included in a "Summary and Update" produced by NOAA, written by the Puget Sound Water Quality Authority and made public this past December. As of this writing, five regional "public information forums" are to be conducted in early January. The public will also have an opportunity to share their own ideas on the proposal (For copies of the Summary and Update, contact Linda Maxson, Sanctuaries and Reserves Division, NOAA, 7600 Sand Point Way, N.E. Seattle, WA 98115).

In the spring, the State is expected to develop and give to NOAA a unified state position on sanctuary issues and options. State support will be important to ensuring that the DEIS/MP is completed and made available for formal public review.

The "Northern Puget Sound Sanctuary Coalition," created by the Center in 1990, is helping to focus the energies of environmental, citizen, and user groups on the sanctuary designation process and demonstrate support for the proposal. CMC is also keeping citizens informed through the Northern Puget Sound Sanctuary Newsletter. For more information on the coalition, please contact coordinator Nancy DeVaux, P.O. Box 1344, Friday Harbor, WA 98250.

In the meantime, CMC members in Washington State are encouraged to write to the Governor and the Department of Ecology in support of designating northern Puget Sound as a national marine sanctuary.

— *Rachel Saunders*

Courtesy of Nancy DeVaux, Friends of the San Juans

Discussion for Marine Mammals... Cute But With a Price

1. Relocating a species requires a lot of thought. Why do you think the government biologists chose San Nicholas Island? What benefits does this island possess that the other ones do not?

2. What government limits are there on the accidental deaths of marine life? Should these limits be more strict and what actions should be taken if they are violated?

3. Discuss what would happen if the predator/prey balance of the otter and urchin was to lean completely to one side. Take in consideration kelp forests, fishing industry.

4. The Coast Guard loses many channel buoys holed by gun shots believed aimed at the seals resting upon them. Can this sort of behavior be sanctioned?

5. Why are people shooting these marine creatures?

6. This is the other side of the controversy. We have fishermen that are trying hard to make a living, to them the sea lion is an adversary. Can you develop a good supporting argument for these people?

7. What are the other reasons, besides fishing industry pressure, that contributes to the decline of the sea lion?

8. Is the Puget Sound a good area for a sanctuary? What sorts of wildlife can be found here?

9. What do you think will be the economic cost of this Puget Sound Sanctuary?

10. Can there ever be a marine mammal plan that everybody can support?

Logging and the New Deal

Michele Garden

In this study the logging practices of the U.S. Forest Service comes under scrutiny. Sometimes the deals arranged by Government services are not in our best interest. In this case you be the judge. Should we stimulate the local economy at a high subsidy or should we protect the resource?

America has an insatiable appetite for convenient, warm, wood-framed homes, fine wood furniture, paper products, packaging and pencils. It is the logging and paper industries that meet this demand. In doing this they provide jobs in felling, assessing, transporting, sawmill operations, and pulpmills, not to mention the various associated manufacturing jobs. The logging industry also contributes sizeable tax revenues. As of 1988, harvesting of federally owned timber generated more than one billion dollars annually in taxes and private timber harvests added another two billion dollars to this amount.

We human animals also have a need for the non-timber resources provided by our forests. Aside from the obvious aesthetic value of a sylvan landscape, forests provide essential wildlife habitats and a balance to fragile ecosystems in the form of a sink for carbon-dioxide and production of oxygen. The nation's National Forests also provide thousands of jobs in fishing, hunting, tourism and land management jobs.

The government agency that must balance these oft-times conflicting needs is the Forest Service. The 29,000 men and women employed by the Forest Service to manage the National Forests must follow the Multiple Use Sustained Yield Act of 1960 which states that the forests are to be managed for a variety of resources – outdoor recreation, range, timber, watershed, and fish and wildlife (and, through other laws, minerals). The National Forest system encompasses 191 million acres of publicly owned forests and grasslands, the bulk of the system (over 165 million acres) being found in 11 western-most contiguous states and Alaska. At least 13% of the timber cut each year in the U.S. and 50% of the nation's primary home-building material, softwood sawtimber, comes from the National Forests.

An excellent example of the connection between the logging industry and the Forest Service can be found in America's largest National Forest, the Tongass in southeastern Alaska. The Tongass is one of the last temperate-zone rainforests in North America and covers 16.9 million acres, 11,000 miles of it running along coastlines and fjords. Of this 16.9 million acres, 5.7 million are commercial quality forest land, known to be among the most productive timberlands in the world. Southeastern Alaska, an area bigger than West Virginia, is home to approximately 65,000 people whose lifestyles and livelihoods are interlinked with the forest. A majority of the jobs are in logging, commercial fishing, and government.

Shortly after World War II, the U.S. government strove to colonize Alaska's southeast panhandle and timber seemed a natural base for this expansion. Although there was a huge and seemingly limitless supply, it was too far from markets and too inaccessible to elicit much response from the logging industries. In order to lure these companies and create year-round employment, the government offered long-term (50 year), special-incentive contracts and two companies signed on. They were Ketchikan Pulp Company (now a Louisiana-Pacific subsidiary) and Alaska Pulp Corporation, a Japanese producer. To this day they are the two major logging corporations in Tongass and remove two-thirds of its annual harvest.

In 1980 Congress passed the Alaska National Interest Lands Conservation Act (ANILCA). It designated one-third of Tongass National Forest as a wilderness area and, in order to maintain timber industry jobs, mandated that the Forest Service must spend $40 million per year to build logging access roads, thin young stands, and conduct research into logging techniques. ANILCA also stated that the Forest Service could lower timber prices to below market levels.

In the mid 1980s environmentalists and other critics became quite vocal in their concerns that the Tongass National Forest was being mismanaged and abused by the government and the logging industries. They claimed the Forest Service was selling timber much too cheaply. A spokesman from the Ketchikan Pulp Mill said his company paid $3.00/1000 board-feet for logs; if taken from the Oregon coast, that much timber would cost $150. The passing of ANILCA in 1980 required the Forest Service to supply 4.5 billion board-feet of timber per decade to be sold. Critics claimed that this was much more than the market demanded in the mid to late 1980s. Environmentalists also argued that the Forest Service was building logging roads into pristine wilderness where, due to decreased demand for timber, logging would never be done. Also of concern was the effect logging would have on the region's most important commodity-based industry – fishing – and fastest growing industry – tourism. Overall, the costs of the Forest Service's Tongass program were so high, and receipts so low, that the government received back only one cent for each dollar spent. From 1980 to 1988 only half of the timber offered for sale was actually purchased. During this same time period, even though the Forest Service spent $300 million dollars to protect the timber industry, jobs declined from 3,000 to 1,750 – a 40% drop.

The logging industries rebuttal to these

accusations was vigorous. Of the 16.9 million acres in the Tongass, only 9 million were covered by forests. Of that 9 million, only 5.7 are commercial forests – wilderness areas covering 4 million of those acres leaves just 1.7 million for the loggers to work. Surely logging this scant 10% of the Tongass National Forest will not harm the overall ecosystem or subsidiary industries. Logging industries stated the job loss would cause considerable economic hardship in southeastern Alaska. An example being the Ketchikan Pulp Mill which employed 520 workers at the mill, 375 in the woods, and contractors employing an additional 465 people – the total being 1,350 workers. The Ketchikan operations also contributed $5 million monthly to the economy of the surrounding area. The Timber Association of California voiced concern about imports vs. exports of logs. As of 1988, the U.S. was a net importer of wood; 12.2 billion board-feet of softwood more imported than exported. This volume represented approximately $2-2.5 billion being contributed to an already unfavorable trade balance. Since the Tongass National Forest was a leader in export of solid wood products, a cutback in logging would mean an even greater trade deficit.

To the logging industries displeasure, Congress passed the Tongass Reform Bill in November of 1990. It eliminated the automatic $40 million annual appropriation for the logging program and the 4.5 billion board-feet per decade target in Tongass. It modified the long-term (50 year) contract with the two logging companies in Tongass. The bill also protected over one million acres of land (300,000 as wilderness areas) as off-limits to logging as well as providing 100 foot wide buffer strips along some Class I and II streams (important to the salmon fishing industry which is now much larger than the timber industry in southeast Alaska). Finally, and most importantly, it directed the Forest Service to supply the local industry with enough timber to meet "market demand" provided that it does so in a manner consistent with the sound management of all the forest's natural resources. Although it met with the logging industries disapproval, the bill is still in effect today.

Below-cost timber sales, prevalent in the Tongass in the 80s, still seem to be a problem for the Forest Service today. Each year the Forest Service makes timber available for sale in the 120 publicly owned National Forests. The Forest Service incurs the cost of building the logging roads and administering the sales. The problem comes not from the profit of $5.1 million made in 22 of the National Forests, but in the staggering loss of $256.8 million in the remaining 98 National Forests. This means that taxpayers are subsidizing a removal of public resources. Some environmental benefits such as returning a forest to a preferred type of tree, improving wildlife habitat, providing access for outdoor recreation, or protecting forests from fire, insects, and tree disease can be said to justify below-cost timber sales, but the profit losses far outweigh any benefits. Many Forest Service critics are saying the government agency is focusing too much attention on serving the logging industry rather than protecting all of the forest's natural resources as the Multiple Use Sustained Yield Act insists that they do. Environmental analysts state that we must create incentives to use materials efficiently – incentives such as high prices for scarce resources currently being sold at subsidized and artificially-low prices.

Another current dilemma is the amount of logging exports. Many people are blaming logging exports for domestic job losses. Currently 90-95% of U.S. softwood log exports shipped from the west coast go to Asian Countries (40% to Japan). Of this amount, 2.6 billion board-feet are from Washington State, 1 billion board-feet are from Oregon, 520 billion board-feet are from Alaska, and 91 million board-feet are from California. Almost all log exports are from privately owned land since the export of logs from federal lands is illegal. Logging industry spokesmen claim that a benefit of exporting is that log and timber prices in western United States are higher than if the log export market didn't exist. This also provides needed jobs and revenue since exporters are willing to pay higher prices than domestic wood processors. For public agencies, this means greater revenue for public projects. For instance, the Washington Department of Natural Resources says a ban on export of logs from state lands would lower revenues by over $100 million a year. A further argument for logging exports is that the approximately $2 billion per year in exporting makes a sizeable dent in the nation's trade deficit, especially with the country of Japan. In terms of job loss, banning log exports would result in some additional sawmill jobs being created, however this would be offset by the jobs lost in the export industry, especially the longshore jobs which pay higher wages than non-union sawmill jobs. Therefore, the net economic effect of further restriction on log exports would be negative.

The Forest Service over the past few years has been attempting to ameliorate differences between logging and environmental interests by practicing a "new forestry" that avoids clear-cutting, but instead leaves some large trees and downed logs intact to make restoration and wildlife recovery more likely. It will be interesting to see if this new attack at an old problem will please all interests concerned.

Reading 1

Questions and Concerns

READING 1

As you read this article consider the following:

1. Are you aware of the facts concerning the sale of Alaskan timber to Japan?
2. Have you been making judgements based upon misinformation regarding the role of the Japanese in the logging controversy?
3. Do you think the facts are correct, is this article a non biased assessment?

LOG EXPORTS: An Update

Emotion and misunderstanding mark this controversial practice. Here's a look at the pros and cons, facts, myths, and prejudices.

By JAY GRUENFELD and BOB FLYNN

At the New Zealand port of Tauranga last February, I watched as radiata pine logs were loaded on a ship headed for Japan. Each log was tagged, and the tag numbers were entered in a book by an attractive lady. Today this is a normal scene. Things have changed from 30 years ago when I was first involved in the log-export business. Then, New Zealand radiata pine logs were rarely exported, and help at an export dock was almost totally male. But one thing has not changed—the controversy over log exports.

Exporting unprocessed logs has been controversial in the United States since the early 1960s. As timber-supply shortages force some mills in the Pacific Northwest to close because of higher log prices, the controversy grows even more rancorous. Are log exports to blame for the timber shortage? Does exporting logs equate to exporting jobs, or are log exports an important part of this nation's timber industry? Here, in addition to explaining some of the characteristics of the log-export industry, we'll discuss some of the arguments for and against allowing this practice to continue.

LOG EXPORTS

Almost 96 percent of the logs exported from the United States in 1988 were softwoods—primarily Douglas-fir (50 percent), western hemlock (31 percent), and sitka spruce (10 percent)—from privately owned, second-growth forests in the Pacific Northwest.

About 90-95 percent of U.S. softwood log exports are shipped from West Coast ports to Asian countries. Washington alone accounts for more than half of the nation's softwood log exports, with roughly 2.6 billion board-feet (BF) in 1988. (By comparison, Oregon exported around one billion BF, Alaska 520 million BF, and California 91 million BF.)

Reading some newspaper accounts of log exports might lead an observer to believe that much of our publicly owned old-growth forests are being clearcut and shipped to Japan as logs. This is simply not the case. Most log exports originate from privately owned land (in fact, export of logs from federal lands is illegal), and with the exception of Alaska, most log exports are from second-growth forests.

In 1980, about 90 percent of U.S. softwood log exports went to Japan. This has changed during the last decade, as China has become a major market for U.S. logs, and South Korea's imports have grown steadily. By 1988, only 52 percent of U.S. softwood log exports went to Japan, while 24 percent went to China, 13 percent to South Korea, and 8 percent to Canada.

Log exports from the eastern United States are a relatively minor part of the total export picture, and most of these are hardwoods such as red or white oak. They are shipped to Western Europe, Japan, Taiwan, and Korea; most of the wood shipped to Taiwan and Korea is used to manufacture furniture for re-export to the United States. Although this furniture trade represents a very small amount of

Courtesy of American Forests

the total log-export market, it appears to have given rise to one of the many myths surrounding log exports. In truth, of the huge volume of coniferous logs exported to Asian countries, only a very small amount comes back as paper or packaging, and a little spruce in musical instruments.

The furniture trade may have given rise to an apocryphal story as well. Someone asked us recently, in all seriousness, if we had heard of a Japanese ship that loaded logs in Washington, processed them into lumber while en route to Japan, and unloaded finished lumber in Tokyo. This same story made the rounds back in the 1960s, but at that time the finished lumber was unloaded back in the United States. In fact, virtually all lumber imports in the United States come from Canada.

NO SHORT-TERM DEAL

In the early 1960s, a couple of Weyerhaeuser Company executives in Tacoma, Washington, advised Jay Gruenfeld on his career path. "Get out of log exports," they said. "It's just a flash-in-the-pan. Get into something with a future."

Well, it turned out they were wrong, and even today, despite pressures to limit log exports, it is a big and usually profitable business for the owners of logs that can be exported.

One reason for the recent rush of news articles on log exports is the record volume being shipped. Log exports from the U.S. West Coast totalled 4.2 billion BF in 1989, down slightly from the record 4.3 billion BF in 1988. For Oregon and Washington, the total figure was 3.7 billion BF. Though this is up about 14 percent from 1979 levels, during the past decade the share of Oregon's and Washington's timber harvest going to the export markets has been remarkably stable. Between 20 and 25 percent of the timber harvest in these two states has been exported during the 1980s.

HIGHER LOG PRICES

Without question, log and timber prices in the western United States are higher than they would be if the log-export market did not exist. Exporters are generally willing to pay significantly higher prices than domestic wood processors are, especially for better-quality logs. Also, the greater overall competition for logs stimulates higher bid prices on both public and private timber.

Whether this is good or bad depends on one's point of view. Private timberland owners benefit from higher prices (and the benefits accrue not only to the Weyerhaeusers but also to the other 65,000 members of the American Tree Farm System and many other timber owners). The higher returns provide an important incentive for private landowners to invest in intensive forest management.

For public agencies, the higher timber prices mean greater revenues for public projects. For example, most of the revenue for timber sold from Washington state lands goes to various trusts, to be used for school construction. The Washington Department of Natural Resources estimates that a ban on the export of logs from state lands would lower their revenues by over $100 million per year.

However, higher timber prices hurt the northwestern lumber and plywood manufacturers. Their primary competitors in the U.S. South and Canada do not have to fight log exporters for their raw materials, and so have a competitive advantage over West Coast producers. Higher raw-material costs mean lower profits, which make it more difficult for western mills to invest in new equipment.

Log prices have become so high in the western U.S. that many manufacturers argue they simply cannot process the wood into lumber and plywood without losing money. These current very high timber prices are not just a result of the demand for log exports, however. Prices are a function of the interaction of demand and supply, and it is the timber supply in the West that is changing most rapidly. The volume of timber sold by the national forests on the west side of Oregon and Washington in fiscal 1989 was only 29 percent of their planned timber-sale program. Since the volume of log exports actually dropped slightly in 1989, most of the recent log price increases are primarily a result of reductions in timber supply, not competition from log exports.

RESTRICTIONS

Currently, logs from federally owned timberlands, including U.S. Forest Service and Bureau of Land Management lands, cannot be exported (with some very minor exceptions). It is also illegal for a mill to purchase federal timber while exporting logs from its own lands in the same geographic area. Some of the recent legislation on log exports seeks to reduce the amount of this "substitution." One Oregon Congressman said recently that the substitution loop-holes were big enough to "drive a log truck through."

Through the early 1980s, Alaska, Oregon, and California prohibited the export of logs harvested from state-owned forests. In 1984, the Supreme Court declared such restrictions unconstitutional, arguing that only Congress had the right to regulate international trade. Several bills have been introduced in Congress this year to restrict the export of logs from state-owned forests, and it appears increasingly likely that some form of restriction will be passed.

Last year a non-binding referendum in Oregon in favor of banning such log exports passed overwhelmingly. A vote on such a referendum in Washington would probably be much closer, as purchasers of state timber have traditionally been allowed to export the logs, and log exports are a much bigger part of the timber industry there than in Oregon.

EFFECTS ON TRADE

Log exports from the United States totaled over $2 billion in 1988, making an important contribution to balancing our nation's trade deficit, especially with Japan. Also, proponents of log exports argue, exporting logs has helped open the door, albeit slowly, to exports of finished wood products. For example, between 1978 and 1988, log exports from Oregon and Washington to Japan dropped 20 percent in volume, while lumber exports from those states to Japan increased 533 percent. (In terms of value, the ratio of log exports to lumber exports was 3 to 1 in 1988; by comparison, the ratio in 1978 was over 18 to 1!)

Perhaps we should be asking, Why does the log trade still exist? If log exports were restricted further, or banned altogether, wouldn't Japan and the other log-importing countries simply replace those logs with purchases of U.S. wood products? This is not likely. There are a number of barriers to increasing wood-product exports, including:

◆ Construction techniques in Japan and the other log-importing countries are quite different from western methods. For example, most housing construction in Japan is still traditional post-and-beam construction, using 1 x 2-meter dimensions (roughly three by six feet). In many cases, wood is custom cut for carpenters for each house by one of the 17,000 small sawmills in Japan.

◆ The distribution and grading systems in Japan are complex and discourage imports of finished wood products. The Department of Commerce, through its "Super 301" trade negotiations, is trying to force the Japanese to open up their markets to more U.S. finished wood products.

◆ Log-importing countries have other options available. If faced with a reduced supply of logs from the United States, they would first try to buy logs from another source, such as the Soviet Union, Chile, or New Zealand. If they do increase lumber imports, they will buy from the supplier that meets their size and grade requirements at the lowest cost. Only some of these additional lumber imports would come from the United States.

The Japanese demand for wood will not go away simply because we reduce our timber harvest or restrict log exports. That demand will shift to other countries where the environmental restrictions on timber harvesting are not nearly as strict as in the United States. The irony, then, is that by reserving excessive amounts of timberland for the spotted owl and other species in this country, we may be hastening the demise of many more species in tropical forests.

◆ Much time, money, and effort have been spent by U.S. producers trying to convince the Japanese to change their housing style to western platform-type construction. This has been partially successful; about five percent of new wooden homes built in Japan in 1988 were U.S. platform type. The companies that have had the most success in marketing finished products to Japan have been those, such as Vanport Manufacturing of Boring, Oregon, which have concentrated on producing to meet Japanese specifications.

In the early 1960s, I questioned some of Weyerhaeuser's log customers in Japan on how we could sell them more finished products instead of logs—for example, veneer and plywood. "First," I was told, "you must understand that in Japan we do not use 4 x 8-foot plywood; we want (roughly) 3 x 6-foot sheets." I brought that message back, and others talked about it and worked on it for years. About 25 years later, Weyerhaeuser finally built the first (and only!) mill to produce 3 x 6-foot veneer for export to Japan.

We simply can't expect the Japanese to willingly switch from buying our logs to buying finished products until we make the size and quality of products that they demand.

EMPLOYMENT

For years, those opposed to log exports have argued that exporting logs meant exporting jobs. Recent economic studies have reached different conclusions regarding the net effect on employment of a log-export ban. In general, banning log exports would result in some additional sawmill jobs being created, but these would be mostly offset by jobs lost in the log-export industry. Moreover, longshore jobs generally pay substantially more than nonunion sawmill jobs, and it seems very possible that the net economic effect of further restrictions on log exports would be negative.

The Washington Department of Natural Resources estimates that any sawmilling jobs created by restricting log exports from state lands would cost the school trusts over $100,000 per job per year, because of lower timber revenues.

It is important to remember that the politics of employment affects decisions in Japan, as well as here, on policies toward wood-products imports. In 1980 Japan had 23,000 sawmills, but by 1989 approximately 6,000 of these had closed.

PROPERTY RIGHTS

An initiative proposed by a timber-industry association in California would, if passed, require the state to petition Congress to ban all log exports from California, including those from private lands. While exports from California represent a small percentage of the total from the West Coast, this is significant because it represents the first serious attempt to prohibit exports from private lands. It is unclear whether private companies would be compensated for loss of revenue.

Such a move in California might have a chilling effect on forest management in other states. How willing would companies be to invest heavily in forest improvement if they think they might be unable to market their timber at the highest price?

CRYSTAL BALL

In a similar article in AMERICAN FORESTS in 1981, Jay Gruenfeld made some predictions concerning log exports for the 1980s that were pretty close to the mark. We'd like to stick out our necks and look ahead to the 1990s:

◆ Exports from the U.S. will probably decrease during the present decade because of a combination of factors, including continued high domestic prices, decreased availability of logs from southeast Alaska, and possibly some form of government restrictions.

◆ The Asian log-importing countries will take a number of steps to offset this decrease in U.S. log exports. Those steps will include increasing purchase of logs from other sources, especially radiata pine from Chile and New Zealand; importing greater quantities of finished wood products, including some from the United States; increasing the domestic timber harvest somewhat in Japan; and possibly using more wood substitutes.

◆ The controversy over log exports will continue. **AF**

Jay Gruenfeld is the president and Bob Flynn is the economist for Jay Gruenfeld Associates, which is a forest-resource and management-consulting firm that is located in Seattle, Washington.

Questions and Concerns

READING 2

As you read this article consider the following:

Are the loggers and the environmentalists ever going to reach an accord?

Turmoil in the Tongass

Our largest National Forest—nearly 17 million acres of wild Southeast Alaska—is aboil with a resource-use controversy that touches every political, economic, and philosophical interest. Stakes are high and solutions hard to come by. Only a change in basic forest policy may help.

By KAREN FRANKLIN

The Ancient Forest

Southeast Alaska is a 17-million-acre stretch of wildness—remote, resplendent, and underappreciated. Embraced by the Tongass National Forest, the country's largest, the narrow 500-mile-long fringe of wild coastline lavishes its flawless gems on nature worshippers and adventure seekers, tourists and timber cutters.

Witness the deep rainforest, its colossal 450-year-old spruce and hemlock, its spongy mosses and lush ferns, its countless streams and waterfalls, its snow-crested peaks. The ancient ecosystem remains intact, an unrivaled scientific utopia that demonstrates forest succession in its purest form—from the lichens that paint the glacier-gnawed rock to the vast climax stands like those that once cloaked the entire Pacific Northwest. Glacial movements sculpted 11,000 miles of shoreline into cliffed fjords, mountain-rimmed cirques, hidden beaches, and forbidding icefields. Some 150 inches of annual precipitation, mostly rain, keeps this land wet, green, and mild all year.

Every fall a late run of salmon attracts the biggest known congregation of bald eagles—and a small throng of camera-laden birders—to the Chilkat River near the top of the Alaska Panhandle. The highest density of grizzlies on earth populates Admiralty Island, a 939,000-acre Wilderness Area, one of countless places on the Tongass where brown bears still flourish. Deer, moose, wolves, seals, sea lions, whales, ravens, and hundreds of other species teem in this forest and its adjacent waters.

The lifestyles and livelihoods of Southeast Alaska's 60,000 residents come predominantly from the forest. Commercial fishermen who live on the edges of the Tongass supply salmon, crab, shrimp, cod, and halibut to local canneries, and pump a substantial amount of money and employment into the economy. Some 90 percent of the local households hunt and fish for sport or subsistence.

Historically, much of the area's economy has revolved around timber, but that trade has fallen dramatically, epitomizing industry-wide trends. Meanwhile, tourists have sdiscovered the Panhandle en masse, bringing essential business to bush

Freelance writer Karen Franklin is the former assistant editor of this magazine.

pilots, hunting and fishing guides, charter boatmen, restaurants, hotels, and shops.

Southeast Alaskans are keeping close tabs on their cherished land and abundant resources, many facing off with the Forest Service in the process. Driven by the Tongass Land Management Plan, they have entrenched the Forest Service in land-use battles on every philosophical, political, scientific, and economic front.

The Battle Lines

In 1979 the Tongass Land Management Plan saw daylight, at least six years ahead of the 122 other forest plans mandated by the National Forest Management Act (NFMA). In the eight years since, embattled planners have faced a deluge of appeals, lawsuits, and Congressional intervention on a scale experienced by no other National Forest. And such action is likely to continue years into the future.

The conflicts largely pivot around provisions in the Alaska National Interest Lands Conservation Act (1980), also known as the Alaska Lands Act or ANILCA, that seem to fly in the face of the ecology-minded National Forest Management Act (1976). According to Forest Service interpretations, ANILCA stipulates a target yield of 4.5 billion board-feet per decade—about 450 million board-feet per year, regardless of market conditions, as laid out by NFMA.

To ensure that the "450" goal can be met, the law also guarantees an annual "timber supply fund," outside the realm of regular appropriations, of *at least* $40 million to build logging roads, prepare harvest sites, and administer the sales of marginal timber stands. No other National Forest's timber program enjoys virtual exemption from yearly Congressional budgeting. In 1982, fund expenditures reached more than $82 million, and sale offerings came from higher-quality stands than ANILCA had predicted.

The provision is explained as a tradeoff for the five million Wilderness acres ANILCA also took out of harvestable lands—a compromise considered necessary to sustain the historic cut levels deemed vital to the local economy. Those Wilderness lands actually contain relatively little high-quality commercial timber, however. "Much of the Wilderness is rock and ice and steep areas," according to Gary Morrison, Alaska Coordinator for the Forest Service. Ninety-one percent of the highest-volume sites on the forest are outside of Wilderness and are still available for sale, says Bob Turnage, resource economist for the Wilderness Society.

Little did Congress know it in 1980, but harvests of historic proportions would quickly become impracticable. Shortly after ANILCA went through, the bottom fell out of the Alaskan timber market, triggering heated controversies about whether it makes sense to continue implementing the law's policies. Environmentalists and scientists have argued that the harvest goal is too ambitious. It puts the ecosystem at risk unnecessarily and adds a substantial deficit to the Treasury, they say.

In fact, taxpayers have lost more than $50 million a year on this timber program. Between 1980 and 1985, for example, ongoing road construction readied some 25 percent of the roadless areas outside of Wilderness for logging, says Wilderness Society attorney Steve Richardson. But chances are the timber won't be cut until the market improves — and according to the agency's own findings, the future looks dim for Alaska's timber sales.

"If the plan keeps grinding away, a substantial number of our finest unprotected fish and wildlife areas will be roaded by 1989," says Steve Kallick, attorney for the Southeast Alaska Conservation Council (SEACC), a group named by Congress to keep tabs on the plan. "The Forest Service agenda is to eliminate the option for Wilderness protection for as much land as possible, as soon as possible," charges Bart Koehler, SEACC's Executive Director.

It is widely accepted among the management participants inside and outside the agency that ANILCA's stipulations have failed to foster industry health—or forest health, environmentalists point out. "Despite massive federal expenditures, the Tongass timber program has not accomplished its stated goal of preserving timber-industry jobs," Koehler says. "It has, however, resulted in the destruction of important fish and wildlife habitat and the renewable resources on which much of Southeast Alaska's economy depends."

But further action to correct the situation is beyond the Forest Service's power, according to land managers. "If the law isn't changed, the agency is duty-bound to follow it," says planner Rai Behnert. "Whether it makes sense to prepare the amount of timber the provision calls for, that's questionable."

The Forest Service could exercise more discretion than it lets on, though, Kallick says, contending that the "450" was meant as a goal, not a mandate. The Congressional Record quotes Congressman Morris Udall, Chairman of the House Interior Committee, as saying on the House floor during the ANILCA debates: "No more timber should be offered for sale than the Forest Service can reasonably expect to sell."

And while environmentalists accuse the agency of supporting failing industry at taxpayers' expense, timber spokesmen have also complained bitterly about forest policy, particularly the way ANILCA has been implemented.

"The Forest Service doesn't admin-ister the timber-supply fund as it should," says Jim Clark, attorney for the Alaska Loggers Association. "They've been building roads where there's the worst timber, so the money hasn't done us any good. It's a scandal—the industry has been absolutely screwed." Timber producers could log more efficiently, he charges, were it not for obstacles imposed by ANILCA and the Forest Service.

Underpinning all those controversies are heated arguments over two 50-year operating contracts. The Forest Service offered them to timber companies some 30 years ago to log on the forest and boost the local economy. Issued long before Congress imposed any environmental protections on Forest management, the contracts profoundly affect planning decisions.

"The contracts are a very big tail wagging the dog," Koehler says. "They control the forest, and that's not how NFMA should work."

Two multi-national timber corporations, one of them wholly owned by a Japanese business consortium, hold the contracts. They market virtually all the wood cut on the Tongass to Japan, raising serious doubts about their benefit to the region or the country. And given market conditions, many locals and observers have challenged the premise that timber's role in the economy is strong enough to justify retaining the contracts and current harvest levels.

Southeast's commercial fishermen, whose healthy, growing industry injects more employment into the economy than does timber, are angry about the plan. "The plan has definitely affected me and my industry," says fisherman Sig Mathisen, President of the Petersburg Vessel Owners Association. "The '450' is a cost to the fishing industry and to the habitat of the area. It is being jeopardized for a timber resource that's being sold to Japan at negligible returns to the country, and the tradeoff hasn't been worthwhile."

State natural-resources agencies support his contention. The Alaska Department of Fish and Game has cited damage to fish habitat resulting from logging, and the Alaska Department of Environmental Conservation has charged inadequate water-quality monitoring.

Tourism is growing faster than any other industry in Southeast Alaska. The local economy could suffer great losses if the timber program hurt the chief draw of travelers—the awesome scenery and outdoor opportunities.

The accusations of unsound public land management and misuse of taxpayers' money would undoubtedly ease off if Alaskan Natives supplied more timber from their private forest holdings. But their efforts to gain a foothold in the pulp market have been frustrated, largely by forest policy.

Though Native corporations own almost half the highest-quality commercial timber in the region, they can't match the below-market rates offered by the Forest Service, according to Bob Turnage of the Wilderness Society. Not only does the agency pick up much of the tab for roading and sale preparation on the forest, in 1981 and 1982 it dropped its stumpage rates by 99 to 999 percent. Spruce went down from $215 per thousand board-feet to $2.26; cedar declined from $1,000 to $1.22.

"As long as the contract-holders can obtain as much timber as they want under such good conditions, they have no incentive to buy from anyone else," Turnage says.

Unable to compete with the Forest Service to sell pulpwood, Natives market primarily round sawlog timber to Japan, a sale technique prohibited by law for wood cut on the Tongass. And because clearcutting has proven the most economical way to cut trees on the forest, they leave the pulp-quality trees lying on the ground, unsold, which puts further pressure on habitat.

Environmentalists don't see stewardship shining through on the Tongass, and they have taken steps to reroute forest policy. SEACC and the Wilderness Society have recommended that the 50-year contracts be canceled and replaced by competitive bidding on short-term sales. They have also asked Congress to repeal both the timer-supply fund and the "450" harvest goal.

"This will not eliminate the timber program from the Tongass, but it will force the Forest Service to be cost-effective and accountable. Responsible timber management will result," says a SEACC report.

The timber industry suspects SEACC's motives, fearing that if those efforts succeed, environmentalists might be just powerful enough to wipe out timber harvesting on the Tongass altogether. "If they've got the muscle to knock out [ANILCA's timber provisions], why would they want timber funded at all?" asks Jim Clark.

But Koehler answers, "All we want is for the Tongass to be treated the same as any other National Forest."

The Wildlife Plan

"For wildlife, old-growth forests are uniquely valuable," says biologist John Schoen of the Alaska Department of Fish and Game (ADF&G).

The Tongass plan calls for clearcutting harvest areas every 100 to 120 years, which for all practical purposes will render the old-growth forests nonrenewable. Permanent habitat changes will inevitably result from the rotation schedule, scientists agree.

The concurrence of 300- to 750-year-old spruce and hemlock with fallen logs, snags, saplings, young trees, evergreen herbs, and lush ferns, plus countless rivers and streams, furnishes unmatched habitat conditions to the forest animals. Both wildlife and timber producers tend to favor the highest-volume stands, originally amounting to some 200,000 acres. About half of that still remains, and 91 percent of it is available for logging, according to Wilderness Society analysis. The resulting second growth will not fulfill the habitat requirements of many wildlife populations.

Sitka blacktail deer, one of many "indicator species" that effectively gauge ecological conditions, are being studied intensively and offer some telling insights on what the timber program could mean for wildlife. ADF&G biologists have predicted significant declines in deer as uneven old growth is converted to even-aged second growth. In winter, particularly during periods of heavy snow, the animals inhabit the old-growth stands almost exclusively. The mature forest's multilayered canopy admits sufficient sunlight for understory plants to grow all year, and keeps enough snow off the ground to make winter forage accessible to deer.

The timber operation will upset that well-laid structure. Though clearcutting produces an initial abundance of browse, for the majority of the rotation period even-aged, uniformly spaced trees create a dense overstory that blocks out too much light to foster understory growth. "Browse production improves immediately for about 25 years, and then declines dramatically for centuries," Schoen says.

The Tongass stands, once cut, will not be afforded the centuries needed to recover. And during the initial succession period when forage burgeons, snows will cover up browse and cause high mortality rates.

Wolves depend primarily on deer for prey. Their populations are expected to decline proportionally with their chief food source. Mountain goats utilize old growth in much the same way as do deer—during snowy winters, they seek the shelter and generous food supply of the mature stands.

Grizzly bears prefer old growth and tend to avoid regenerating clearcuts. They are known to circle widely around human activity, and the plan's roadbuilding program is likely to reduce the space and seclusion that this last thriving grizzly population needs.

The Forest Service—acknowledging that replacing mature stands with second growth will undermine habitat—is trying to mitigate the damage without changing the slated harvest level. Agency managers have designed several programs with that goal.

One of them involves clearcutting in a mosaic pattern to leave patches of old growth between the harvest areas, and treating the clearcuts to optimize forage conditions. "What we'll really have is the full range of successional stages," says Phil Janik, Director of Wildlife and Fisheries Management for the Alaska region.

Whether that strategy works will depend largely on how the Forest Service selects the old growth to retain between the harvest units, and questions about how much to preserve or where to leave it have not been fully answered. And ADF&G biologists believe that even-aged patches of second growth degrade habitat, regardless of their juxtaposition to mature stands.

As the second-growth forest evolves, the stands will be thinned to create openings, admitting light and stimulating understory growth. But the advantages of such management remain speculative—indications are that it won't offer the unique habitat conditions supplied by old growth.

"Second-growth management works to some degree," says Janik, "but I'd quickly qualify that by saying we're not trying to recreate old growth; we're trying to maximize forage conditions in a cut-over area."

The plan's potential impact on fish habitat, too, has stirred a lot of attention and worry. Thousands of streams that enlace Southeast Alaska clearly and quickly reflect environmental changes. Scientists have cited many stream conditions that can shift as a result of logging, which can trigger calamity for the forest's vital fisheries.

In response, the Forest Service has employed some management prescriptions to mitigate the effects of logging and improve habitat. Millions of dollars have been spent on fish enhancement, and record catches have ensued, according to the agency. Land managers anticipate an increase of some six million fish as a result of such efforts and thanks also to ADF&G fisheries management and favorable climatic conditions.

But despite the hopeful outlook, the improvement programs fall far short of counterbalancing the harmful effects of logging, says ADF&G fisheries biologist Gary Gunstrom. "The Forest Service puts about $2 million a year into habitat enhancement and makes a big deal out of that," he says. "But at the same time, it puts $50 million or $60 million into logging operations. In every case, the value of the fish by far exceeds the value of the logs, especially when you consider that no one wants to buy the logs anyway."

So far the Forest Service has not adequately monitored changes in stream conditions after harvesting, say both professional observers and agency managers. Degraded fish habitat is difficult to quantify, but biologists maintain that lack of measurements does not mean problems don't exist.

"No one has been looking to see if water quality is intact throughout Southeast Alaska," says Bill Leitch, ecologist for the Alaska Department of Environmental Conservation. "It's not possible to log here as it's being done without some long-term effects."

Janik concurs that stream monitoring could be improved. The Forest Service is looking at changes in spawning and rearing habitat that occur after logging, he says, "but not to the degree we'd like to get to."

These debates are nothing new on the Tongass—or on any National Forest. But in the lower 48 states, it's a fact of life that only small parcels of pure wildland are left; not so in Alaska. Much of the Tongass' vast backcountry has not even been explored yet. Researchers know relatively little about the ecosystem dynamics of the virgin forest tracts, knowledge that could both guide land management and contribute immeasurably to science.

"Scientists are trying desperately to learn what they can in the little time remaining," Schoen says. "The problem is not with the known, but with the unknown. Our final opportunity is in Southeast Alaska."

The Tongass Tomorrow

Environmentalists, timber producers, scientists, and bureaucrats agree about one thing regarding the Tongass: in one way or another, its management isn't working. But many people are laboring hard to correct the problems, and the machinery is cranking in both Southeast Alaska and Washington.

In keeping with the National Forest Management Act, an overhaul of the plan is due in 1989. Preceding the NFMA regulations that govern forest planning, the Tongass plan has never complied procedurally with the law, and the revision should bring the plan's terminology into conformity with national standards. Substantive changes, such as cutting levels, are not anticipated unless Congress removes the timber-supply provisions from the Alaska Lands Act.

To get over that hump, conservationists are lobbying for passage of the Tongass Timber Reform Act, a bill introduced by Robert Mrazek (NY) in the House and by William Proxmire (WI) in the Senate earlier this year.

Should the bill become law, it will: 1) finance the timber program by regular appropriations rather than by a yearly guaranteed fund of at least $40 million, 2) repeal the slated timber-supply goal of 4.5 billion board-feet per decade, 3) reinstate an NFMA provision, currently exempt on the Tongass, that would require the Forest Service to consider economic feasibility of harvesting there, and 4) require the Forest Service to analyze the timber program's impact on the local economy—specifically commercial fishing, tourism, and subsistence. Environmentalists are trying to amend the bill to further force the Forest Service to terminate, modify, or buy back the two long-term timber contracts that highly influence decisions on the forest.

Lobbyists for the timber industry are urging Congress to leave present policies alone, citing the Wilderness Areas that were set aside in exchange for current provisions. "We want ANILCA to stay as is," says Kirsten Held, lobbyist for the Alaska Loggers Association. "It was a compromise, and there's no reason to change it."

"A compromise we never agreed to," Bart Koehler calls it.

The loud, long controversies have attracted the attention of government. The House Appropriations Committee, for example, completed an inquiry into Tongass policy early this year. Congress's General Accounting Office is currently doing its own investigation, and the Department of Agriculture's internal Office of Inspector General has scheduled a study. It's too early to tell how or if those actions will affect future Tongass management. **AF**

Questions and Concerns

Reading 3

READING 3

As you read this article consider the following:

Is the U.S.F.S trying to change its image? What are the issues involved and is the government being candid in its responses to the environmentalist organizations?

NEW FORESTRY NEW HYPE?

Forest Service ecologist Art McKee seems a little sheepish as we gaze across the canyon, where an ancient forest in the heart of the Oregon Cascades was clearcut more than a decade ago. The hillside resembles a Christmas-tree farm, its Douglas firs as uniform as a battalion of recruits. ■ Farther up this fork of Tidbits Creek is a more recent cut that shows how far the science of forestry has come. Clusters of live trees dot the site, cut logs are strewn on the ground, and several snags—dead trees—are still standing. The idea, McKee explains, is to offer varied habitats for wildlife: Birds nest in the snags, small mammals den under logs, and insects and fungi chow down on the dead wood. With his help, I imagine the live trees towering over the seedlings that will be planted next year. ■ This part of Willamette National Forest offers a showcase example of New Forestry—an approach to logging that aims to preserve ecological processes, plant and animal species, and soil productivity. Its techniques were developed in the Pacific Northwest, where dwindling primeval forests are home to more than 100 species of wildlife—including the osprey, the tailed frog, and the northern spotted owl—that depend on the groves. ■ New Forestry is the keystone of a larger initiative within the Forest Service called New Perspecitves. "New Perspectives first focuses on providing for a healthy, vibrant ecosystem, and then delivers sustainable products in the process of maintaining that healthy state," says Charles "Chip" Cartwright, assistant director of the program. According to Cartwright, Forest Service Chief F. Dale Robertson adopted the new approach in 1990 after surveying the long

The drive to turn trees into two-by-fours is sabotaging scientists' best-laid plans.

list of conflicts the agency was facing—litigation, an increasingly militant public, and a Congress simultaneously demanding forest protection and timber. Where the agency attempted to balance logging with diligent protection of the rest of the forest's resources, he noted, it encountered less hostility and was sued less often.

Working in the gray area between preservation and all-out logging, New Forestry was devised to satisfy demands both for healthy forests and for an adequate wood supply. That is precisely its seductive appeal—and its gravest threat. "New Forestry is dangerous because it tells politicians they can have their owls and their timber, too," says Jeffrey St. Clair, former editor of *Forest Watch* magazine. "That's a great message if it's true, a risky one if it isn't."

New Forestry requires managers to think in terms of whole watersheds and landscapes—hundreds of thousands of acres, not just the dozens in a typical logging unit. Its goal is management that mimics nature, allowing enough time for the placid progression of growth, decay, and rebirth to occur, even as it simulates natural disturbances. In the Cascades, the most significant of these is fire. Although a chainsaw can never replicate the soils and landscape left by an erratic thousand-degree blaze, it can leave the forest in a similar condition. The number of cut trees can approximate the amount that would have been burned, and live trees, downed logs, and snags can be left in post-fire proportions.

Across the creek from where McKee and I stand, a ragged carpet of trees covers a hill, evidence of fires that spread across the land in decades gone by, creeping through gullies, charging in fingers up hillsides, and roaring over ridgetops. As a result of these blazes, high places are brushy and young, large trees populate the stream bottoms, and the mid-slope is a mosaic of young and old.

McKee says that researchers have studied scars on tree rings to compile an 800-year history of the frequency and intensity of fires on 150,000 acres in the Cascades. The patterns that emerged serve as guidelines for logging: If 10 percent of a valley used to burn every 50 years or so, an equivalent 10 percent might be loggable over the same period. The size of severely burned patches indicates how large an area to log. For New Forestry advocates, these natural patterns are a road map of the way the forest should be managed.

McKee steers my attention back to a logging site up the creek. "This approximates what you'd find after a moderately severe fire," he says. Not all the trees would be killed by the blaze, and much wood would be left on the ground. The remnant trees and debris are "biological legacies," an inheritance that will allow the stand to resemble old growth within 60 or 70 years, instead of the 200 or more that would be required after a 1970s-era clearcut. The foresters have even considered how long the snags will last and whether enough old trees will be left to create new snags when the first ones topple.

New Forestry has run into its share of flak for trying to steer a middle course between environmentalists and the logging industry. "The industry says 'That's not a clearcut,' and the environmentalists say 'That's not a forest,'" observes Fred Swanson of the agency's Forest Sciences Laboratory in Corvallis, Oregon. According to some wags, New Forestry and the rest of the New Perspectives program, including New Technology and New Alliances, will likely be carried out with New Chainsaws, allowing loggers to buy New Pickup Trucks.

These widely diverging opinions aren't just a matter of blind men describing an elephant—they also reflect differences between the theory and the practice of New Forestry. While scientists earn praise for probing and mimicking the ecological processes of a native forest, there's many a flaw 'twixt the thought and the saw.

The Forest Service faces pressure from Capitol Hill, industry, and factions within its own ranks to extract unsustainable quantities of timber and other resources from public lands. In the past, even when tempered somewhat by environmentalists, these demands have wreaked havoc on national forests. They could easily negate the gains that New Forestry promises.

The gulf between words and deeds is nowhere deeper than in the Shasta Costa basin of Siskiyou National Forest in southwestern Oregon, where the Forest Service recruited local environmentalists as well as industry representatives to help create a model forestry plan for a mostly roadless 23,000-acre watershed. The activists responded favorably: "We wanted to put an end to the war in the woods," recalls Jim Britell of the Kalmiopsis Audubon Society. "No one was enthusiastic about logging in a roadless area, but Oregon was cutting record numbers of trees and would continue to do so until the laws were changed. We figured we'd better come up with a method that didn't trash the place."

The Forest Service wanted to "keep the decisions out of the courtrooms and in the forests where they belong," says silviculturalist Kurt Wiedenmann, leader of the Forest Service's Shasta Costa team through much of the planning process. It also hoped to regain credibility with environmental activists, who had come to distrust the agency after seeing too many devastating clearcuts on public land.

When planning for Shasta Costa began in 1987, cautious optimism, openness, and cooperation carried the day. In the past the agency might have heard comments on the scope of the proposed project, hibernated for a year, produced a thick tome for public comment, and then been surprised when people felt left out. This time, however, the agency involved the public throughout the process, actively soliciting ideas and keeping interested parties abreast of the project through newsletters, the media, and frequent meetings. When the draft environmental impact statement was issued, the full-court press continued with a lavish, 44-page, four-color brochure on Shasta Costa and the entire New Perspectives program.

What emerged was a highly restrained logging plan. Because Shasta Costa lies between two wilderness areas, the Wild Rogue and the Kalmiopsis, foresters mapped corridors to allow animals such as the pine marten and pileated woodpecker to migrate between the two. The Forest Service admitted that "it was time to restore the old-growth resource to more natural levels," in the words of its brochure. It sought to avoid fragmenting the large blocks of ancient forest that already existed, and to preserve the interior habitat required by species such as the northern spotted owl.

The plan for Shasta Costa was nobody's favorite, but it was nothing like what Audubon's Britell calls "the log-your-brains-out alternative." The agency proposed to concentrate logging in younger stands so as to leave the old growth mostly untouched, to construct only two-and-a-half miles of new road, and to cut 25 percent less timber than had been called for under earlier schemes. Ancient forest slated for logging would be treated with New Forestry techniques to retain many of the old trees.

Environmentalists who took part in the process felt they could live with the plan, which was released in the summer of 1990. "If logging has to be done in a roadless area, this is how we would like to see it done," wrote the local Audubon Society in its newsletter. "Shasta Costa gives us a model."

A year later, the coalition had fallen apart. Many of the environmentally benign timber sales turned out to be impractical when inspected on the ground because the trees were too small to be economically attractive. To make up the missing timber volume, Wiedenmann and his colleagues added more old growth to the cut, doubled the planned road construction to five miles, and upped the total harvest by about 20 percent.

These changes went a long way toward restoring the cynicism of the environmentalists. "Agencies are willing to give us an endless supply of elk-viewing areas, maps, rides in helicopters, new trails, and schmoozing," wrote Britell. "Anything and everything except trees."

Deepening activists' dismay was ambiguity about the proposed cuts. For more than a year after the agency released the draft plan, the question of how many trees per acre would be spared in the New Forestry timber sales was left open. "I'm interested in promoting alternative forestry—as soon as I find out what it is," complained activist Julie Norman.

Forest Service personnel might have been able to allay these fears had they not been busy transferring to other districts. The planning phase outlasted three district rangers, three forest supervisors, and three team leaders. "They could replace me with a timber beast who would throw out all the relationships we've built with environmental groups," Wiedenmann told me last May. Two months later, he was transferred to a forest in eastern Oregon.

What most bothered local activists was the fact that Siskiyou National Forest's ten-year blueprint demands 32 million board-feet from Shasta Costa by 1998. Since the three-year Shasta Costa plan would provide only 14 million, another 18 million would have to be extracted in the rest of the decade. The ten-year plan even went so far as to map where further cutting could take place. This enraged environmentalists, who saw the carefully protected heart of Shasta Costa ripped out by future timber sales.

Britell and others propose a straight-forward solution: lower the cut. But right now—especially in light of recent directives from Washington (see "The August Coup," January/February)—there is strong pressure on all national forests to "get the cut out."

"As you go further out in time, you can't hang onto goals like avoiding fragmentation," new team leader Rod Stewart admits. When pressed, he acknowledges that the main advantage of a New Forestry plan for Shasta Costa is that it buys three more years before the remaining old growth is split apart. The moral is clear: Unless New Forestry is paired with a slash in logging, it offers no long-term benefit.

New Forestry also makes little sense unless it is combined with sound management of a forest's other resources. In Lewis and Clark National Forest in Montana, a New Forestry plan calls for cutting small patches of trees that shade and stunt the growth of buffalo berries—a prized grizzly food source. "We no longer harvest timber for timber's sake," says Forest Service wildlife biologist Seth Diamond. "This project is driven by grizzlies."

That sounds like good news indeed, but there's a hitch never mentioned in Forest Service press releases. This particular national forest doesn't have much merchantable timber: What it has instead is oil-and-gas potential, and Lewis and Clark administrators have leased huge swaths of prime grizzly habitat for energy development. Here, the forest and its associated wildlife are bound to lose in the end, with or without benefit of New

Forestry.

These new techniques seem to have been tortured beyond recognition in the "alternatives to clearcutting" being tried in Arkansas' Ouachita National Forest. "It's just the same old clearcutting with a little gloss," says beth Johnson, the Sierra Club's Southern Plains representative. "New Perspectives? We call it new PR."

ONE GREAT FEAR is that New Forestry will be used as an excuse to cut old growth in roadless areas. With 95 percent of the nation's woodlands already logged, critics say, the remaining virgin forests should be protected, not vivisected. Old growth is the only reliable "greenprint" of how a healthy forest functions.

Former Shasta Costa team leader Wiedenmann responds to such concerns by saying, "People forget how resilient these ecosystems are."

"But where does that resilience end?" I ask him, as we stand in the forest he proposes to manage with New Forestry techniques. "Is it impaired when just 9 percent of the old growth is left? Three percent? One percent?"

"We don't know," he admits.

Because New Forestry is still in the experimental stages, the issue of whether to use it in old-growth forests comes down to a choice between humility and hubris. Are we feeling lucky this year? At a recent conference, Forest Service ecologist Michael Amaranthus joked, "You know what strikes the most fear in me? It's when my five-year-old walks toward my computer with a toolbox. That's what we've been like in the forest for 30 years."

But even those who are optimistic about the technical promise of New Forestry have to face that other, even thornier question: Will New Forestry continue to be warped in the name of political expediency? Even the best-laid plans to promote ecological health will go awry if extravagant demands for raw materials are placed on the public forests. Until the current pressure is reduced, giving these lands and their caretakers a bit of breathing room, neither enlightened science nor good intentions can offer much hope.

SETH ZUCKERMAN *lives on the north coast of California. He authored* Saving Our Ancient Forests *(Living Planet, 1991).*

Discussion for Logging and the New Deal

1. What is the full responsibility of our U.S. Forest Service in the Tongass area and do you think they are fulfilling their obligation?

2. What is the 'true cost' of a tree?

3. Stop the fires but estimate the number of trees 'saved' and then log them...is this a workable hypothesis?

4. Logging new growth but leaving old growth – does that make much sense?

5. Just how resilient is an old growth forest?

Introduction

Matt Atwood

The Exxon Valdez Oil Spill

On Good Friday, March 24, 1989, the fully loaded oil tanker, Exxon Valdez, ran aground on Bligh Reef spilling 232,000 barrels (11 million gallons) of crude oil into the pristine beauty of Prince William Sound, Alaska. The oil spread in a Southwesterly direction into the gulf of Alaska covering over 25,000 kilometers of coastal and offshore waters occupied by various species of plants and animals. The Exxon Valdez spill was not the largest spill in the world, but because of the dense population of animals inhabiting the sound, it quickly became one of the most devastating man made disasters in the history of the United States.

Crude oil is a blend of four groups of chemical compounds consisting of aromatic hydrocarbons, aliphatic, sulphurated, and polar compounds. The first few days of a spill are typically the most hazardous because toxic hydrocarbons, such as benzene and toluene, are in their highest concentrations and need a couple of days to weather or evaporate. This is one reason why the most environmental damage is done within the first 10 to 36 hours after an oil spill.

Another element added to the danger of crude oil is the damage inflicted to animals that are in physical contact with it. Since oil is less dense with water, it floats on the waters surface making it accessible to all of the creatures that live and feed in the top water column. Damage inflicted to animals exposed to crude oil range from suffocation or ingestion to exposure of winter elements caused from damage to insulation. Needless to say, animals that were contaminated by direct exposure to crude oil in the first few days of the spill suffered the brunt of the wildlife death toll by a combination of the toxic compound exposure and damage by physical contact.

ENVIRONMENTAL DAMAGE

Both terrestrial and marine animals and plants were affected in some way by the spill but most of the scientific, ecological, and social concern was directed to organisms that would suffer the worst from it. Immediate social and political attention was directed toward the multi million dollar commercial fishing industry in the sound and its affects, if any, to the toxicity of fish caught. Most of the scientific focus went to the marine wildlife of the area with special interest toward animals in the most immediate danger to oil exposure.

EFFECTS ON COMMERCIAL FISHING

In all of Alaska, commercial fishing is the second largest revenue producing industry (behind the oil industry) consisting of pen-reared shellfish (oysters, shrimp, and crabs). Obviously, this multi million dollar industry was given high priority for protection. Exxon studies indicate that free swimming open water adult fish are rarely at risk from oil spills because fish simply swim away from floating oil. However, larvae and eggs are much more vulnerable to developmental death or eventual abnormalities caused by exposure to oil. This observation lead to a question that only time could answer; how much would the oil spill affect the commercial fish population? Major concern was focussed on two of the three commercially important types of fish, salmon and herring. It was believed that since halibut are bottom dwelling fish, their life cycle would not be affected.

SALMON

There are four species of Pacific salmon known to breed in the Prince William Sound area, but most of the concern was focussed on Pink Salmon. The reason is that pink salmon have a two year life cycle (compared to four year cycle of other salmon) and spawn in gravel at entrances of creeks or in intertidal areas where oil damage would be the worst. When oil reached the shore in 1989, young pink salmon were still spawning giving worries to fisherman about possible population impact for the 1990 season. This worry lead to the closing of 2 of the 11 salmon fishing areas due to fear of contaminated fish and because of interference to oil clean up.

Evidently the following 1990 pink salmon catch relieved all worries by breaking all previous pink salmon catching records (over 44 million fish caught compared to previous 29 million catch record).

HERRING

The majority of immediate concern was given to the affects of the oil to the herring population. Herring congregate in huge schools every April, which in 1989, was two weeks after the oil spill. Fishermen feared that the eggs and developing young would be damaged thus severely depleting future catch loads. This fear lead to the closing of the entire herring fishing haul for 1989 (a 12 million dollar loss). The following fishing season would determine the actual affects of oil to herring population.

Within 20 minutes of the opening of the 1990 fishing season, it was over; but the fisherman weren't depressed, they were ecstatic. The entire years seine harvest quota was filled in less than 20 minutes equaling 8,300 tons of fish (32 percent higher than the previous year). Why would there be such an extraordinary abundance of fish? It turns out that there was a record breaking plankton boom in 1989 that supplied food for pink salmon harvested in 1990. The large pink salmon catch could also be attributed to increased hatchery releases. The 1990 herring harvests obviously included fish that would have been caught in the previously closed season of 1989.

Evidence shows that fish were not affected by toxic hydrocarbons. Hydrocarbons are not like DDT or other toxins which accumulate over time through a food chain. Fish have the capacity to metabolize hydrocarbons and expel them from their systems. However, reports of fish being tough and rancid were also reported.

EFFECTS OF OIL ON BIRDS OF THE SOUND

Seabirds were by far the most obvious casualties from the Exxon Valdez spill. It has been estimated that the number of seabirds affected by the spill make up 20% of the entire state of Alaska's total bird population. Enormous seabird colonies, mostly Alcids (murres, puffins, auklets, guillemots, and murrelets), inhabit the Prince William Sound area. The extent to how the oil affected different species relied heavily on feeding behaviors of the species involved. Estimated total deaths range from 100,000-300,000 birds and could have been more if the spill occurred during Summer or Autumn when the highest populations of birds occurs.

Birds that swim on the surface of the water and dive for fish such as Alcids, Loons, and Grebes, were most directly affected by the oil in two ways. First, exposure to toxic hydrocarbons that were present the first few days of the spill have been found to cause serious internal hemorrhaging which eventually leads to death. The other effect was exposure of oil to feathers. Since these birds swim on the surface, the oil (which is also on the surface) causes feathers to stick together which does two things:

1) It traps the bird on the water by inhibiting or restricting its capability to fly, and
2) Cripples the bird's insulating capabilities resulting in death by hypothermia. Ingestion of oil during feather cleaning is also known to have toxic effects on the liver, spleen, and kidneys.

Most of America's bald eagle population can be found in Alaska with an estimated 5,000 eagles in Prince William Sound alone. In fact, it's theorized that since the population of Prince William Sound is so saturated, that the eagles have acquired the ability to suppress typical territorial behavior. This would explain why thousands of eagles have been known to congregate on salmon rich rivers at certain times of the year. Unfortunately, this behavior has also been observed when the similar attraction of many oil drenched carcasses appeared on the shores after the spill. This smorgasbord of carrion has eagle specialists worried. Ingestion of such large quantities of oil threatened to harm developing embryos during this years reproductive reason. According to Audubon Society researchers, one microliter of crude oil on a bald eagles egg can prevent an embryo from maturing.

By mid July, 1989, 153 eagles had been found dead. Much worse, a survey showed that of 330 nests in sample areas, only 111 showed even remote signs of nesting behavior. One week after what should have been peak breeding time, 54 of those nests were void of any eagle activity. Even worse, all 18 nests on Eleanor Island (most heavily oiled island) were completely barren.

Despite these facts, Exxon researchers still believe that the low productive rate of eagles in 1989 was due to disturbances caused by cleanup activities and not by the oil itself. They believe the reason for 1990's active nest count is due to "reduced clean up efforts" and the fact that "eagles do not breed until they are 5-6 years old, any losses among breeding population (adults) are filled by new recruits, at least by the next breeding season." In other words, they believe that only sexually mature eagles scavenge on oiled carcasses and when they die, they are "replaced" by new eagles "waiting" for their breeding territory.

Long term effects are uncertain but several facts are known that may cause serious damage to the rebuilding of populations. Oiled adults unintentionally smear their eggs with oil and egg shells are permeable to oil which is toxic to embryos. Another study shows that exposure to oil can lower the productivity of adults, they simply fail to produce eggs. A major concern consists of a question: Exactly how many breeding adults were actually lost? Ornithologists estimate that only 1 in every 4 dead birds is recovered, that would make a total of 35,279 x 4 = 141,116 birds killed.

Ornithologists believe the devastation to populations of various bird species may take 5-25 years to replace. Exxon, however, is being optimistic by stating the overall "populations of reserved non-breeding adults outside the breeding colonies . . . will replace any losses from the breeding colonies." Recent studies do indicate that both marine bird and eagle population are doing better than expected.

EFFECTS OF OIL ON MARINE MAMMALS

Many unknown variables came to play in the affects of the spill on marine mammals. Only one species, the sea otter, was in serious immediate danger, but what about the 9 other species of marine mammals that inhabit the sound? Would they be damaged from the oil? The answers to their questions, except for a few studies, remain speculative. No marine mammals, besides otters and a few seal pups, were actually recovered and labeled as an "oiled casualty." When marine mammals die, they sink to the ocean floor where the flesh is eaten by invertebrates and only bones show up on shore. This would explain why no cetaceans (whales and dolphins) or pinnepeds (seals and sea lions) were actually recovered.

SEA OTTERS

Prince William Sound hosts a population of an estimated 8,000-10,000 (and even up to Exxon's estimated 16,000) sea otters prior to the spill. In 1911, 50 individuals were placed in the sound in hopes to protect the otter from extinction. Obviously, the otters are resilient little characters. However, after the first few months of the spill, sea otters caught in the oil suffered horribly. While cleaning up and capturing oiled wildlife, 878 sea otter carcasses were recovered with many more undoubtedly not found. Experts estimated that for every 1 otter found, 5 had died making an estimated total of (878 x 5) 4,390 otter deaths resulting from the spill. Even at the modest Exxon estimate of 16,000 otters in the sound prior to the spill 4,390 otter deaths is around 30 percent of the total population. Exxon reports consider these deaths to be "only a small fraction of the otter population." Because of the otters photo-genic qualities and its vulnerability to the oil spill, special rehabilitation privileges were given to otter population damages.

Crude oil affects otters in several ways, both externally and internally. Otters spend a majority of their lives floating on the surface of the water where they can be seen eating shellfish or constantly grooming their incredibly dense fur. Since otters do not have a layer of blubber surrounding their bodies like seals do, they rely on the insulating power of their fur and their high metabolism to keep them warm. These two otter characteristics proved to be damaging to the otters when exposed to crude oil. The affects of crude oil on the otters pelt acts in the same manner and results in the similar symptoms encountered by birds. When the oil penetrates the otters fur, it causes the fur to mat together, thus lowering the pelts effectiveness to thermal-insulate the animal which leads to a chain of other problems. The animal immediately attempts to clean its fur to protect it from hypothermia and to maintain buoyancy in the water. This futile attempt to clean its fur creates another problem of oil ingestion which leads to poison of the kidneys and liver, some had damaged spleens causing internal bleeding. The otters will also attempt to keep warm by increasing its food consumption to increase its metabolic heat, but the animal becomes emaciated and stressed and eventually dies anyway.

Depending on date of exposure, many otters were exposed to the oil when high levels of toxic hydrocarbons where present. This exposure to petroleum hydrocarbons contributed to another set of health problems for the otters. Many deaths were diagnosed as being caused from pulmonary emphysema, anemia, and hypoglycemia. Many otters also had nosebleeds, which suggest that their nasal membranes were also damaged. Stress from capture could produce some of those symptoms as well. The overall intention of Exxon's rehabilitation project was commendable but the future efforts to rehabilitate and conserve populations is unrealistic because it is too expensive and ineffective.

The current evaluation of sea otter population suggests that the majority of the effects of the spill were exclusively short term because of the effective clean up procedures. Populations are already being recorded as stable and growing steadily. Blood parameters and reproductive status have returned to pre-spill standards faster than anticipated. The future of the otter seems better than expected.

EFFECTS ON OTHER MARINE MAMMALS

Marine mammal casualties are, at best, speculative because no bodies were recovered. Since marine mammal carcasses sink, there is no true scientific fact of how many, if any, animals were affected. Since seals and whales rely on blubber for insulation, there is no concern about external damage to these animals except the characteristic "black cap" on some seals heads from surfacing for air through the oil covered surface. This cap indicates that the only danger to seals and whales would have been exposure to toxic hydrocarbons which depends on date and quantity of possible exposure.

Permission was given to sacrifice a few seals in the worst hit areas for autopsy to see the affects, if any, of hydrocarbon exposure. It turns out that some seals suffered from eye problems, such as conjunctivitis, and some seemed unusually lethargic as a result of brain lesions from inhaling the toxic fumes, much like a human brain when "sniffing glue." Traces of damage to the nasal membranes and lungs were also evident. It is even more difficult to determine the impact of the spill on whales. There was a pod of resident killer whales in the sound and according to one specialist, several are reported missing. The long term contamination effects of marine mammals may never be determined. It has also been suggested that toluene, one of the toxic chemicals, acted as an attractant to pinnepeds and otters which drew them to the most dangerous parts of the spill.

EFFECTS ON LAND ANIMALS

Land animals that live in nearby weeds and rivers that lead down to the oiled beaches have not escaped contamination, although they have received little attention. Most of the focus was on the water. Few bodies are liable to be discovered for the simple fact that land animals that feed on the beach return back to the forest. However, 2 black tailed deer have been found dead in tidepools when feeding on contaminated sea grass during low tide. Another casualty was recorded that had nothing to do with oil exposure, but during the oil clean up, a persistent bear was shot when interfering with clean up crews. Bears, foxes, wolverines, and river otters have also been sighted in taking advantage of oiled carcasses washed ashore.

OTHER ENVIRONMENTAL IMPACTS OF OIL EXPOSURE

Invertebrates, such as mollusks, insects, plankton (zooplankton and phytoplankton), crabs and shrimp were also affected by the spill. Particular attention was paid to mollusks and other shellfish in the intertidal areas. Contamination of the majority of intertidal zones were eminent, but resiliency is a factor well known of intertidal organisms. If these animals survived the oil itself, during the cleanup, Exxon crews used a technique for stubborn oil which called for high pressure water at high temperatures which essentially "cooked" these animals at 145 degrees. In terms of abundance and diversity of life, oiled beaches that were left untreated are now similar in most instances to sites where no oil had come ashore. Treated beaches are clearly in worse shape and almost completely sterilized of cold water intertidal organisms. Unfortunately the lesson to be learned by overly ambitious oil spill cleanups is a sensitive one. One expert is quoted as saying, "Sometimes the best, and ironically the most difficult, thing to do in the face of an ecological disaster is to do nothing."

EFFECTS OF OIL ON HUMANS

In the days following the Exxon's spill containment handlers worked for days on end in the same oil soaked clothes. It has already been pointed out that crude oil and mammals don't mix, and sea otters were not the only mammals bathed in crude. Just how dangerous is oil? Skin contact can cause acute dermatitis, not to mention possibly breathing in fumes leading to severe emphysema, as well as liver, kidney, intestinal, adrenal, and bone marrow abnormalities. Luckily, toxic compounds were virtually nonexistent by the time the entire cleanup crew arrived.

Another concern was expressed about the 100 or so native Alaskans who live in oil drenched areas and rely on seafood for 80 percent of their diet. If the fish were contaminated, it would put the natives at a high risk of cancer or other organ damage. No one will really know the extent of human damage until more in depth research is possible.

OIL CONTAINMENT AND CLEAN-UP

Prince William Sound is a part of Alaska that is virtually cut off from the rest of the world. A road system was virtually nonexistent resulting in deployment of all shoreline treatment equipment and personnel by boat. In the early hours of the spill, detergent dispersants were being considered as an early remedy to stop the oil from polluting the sound, but the delay in response made that consideration an impossibility because the spill quickly spread beyond the capabilities of the detergents. Since the sound was so heavily populated with wildlife, using detergents at such a late stage could do more harm than good.

The next obvious step was containment, but the damage of the first few weeks of the spill exposed the incapability and unpreparedness of Exxon to deal with such a large "highly unlikely" spill. Booms of all shapes and sizes were used, but winter storms hit the sound days after the spill forcing the oil up and over the booms enabling the slick to spread uncontrollably. In fact it wasn't until 100 days after the spill that the maximum amount of booms were deployed. By that time, the spread and damage had been done. The only thing the clean up crews could do now was to section bays and stream ways off with booms and clean up oil deposits that were isolated inside section areas.

Oil on the surface of the water was isolated and skimmers (the major oil recovery device used in Exxon's cleanup response) were used to pick up oil floating on the surface by use of a pump that virtually sucks the oil off the surface. The recovered oil is transferred to a receiving tank by use of vacuum equipment and then shipped to a holding vessel for disposal. Most free-oil pick-up (oil floating in the water) was completed by June, 3 months after the spill. Skimmers were also used for shoreline pick up. Booms were laid just off shore to keep oil confined in the cleaning area while on shore pressure fire hoses sprayed the oil off the shore and back into the water for recovery.

The vast majority of oil impacts were located in the intertidal zone (beach area between high and low tide). The lower zone, where flora and fauna are typically prevalent, was the most environmentally sensitive. The approach was to remove, contain and collect the oil from the shoreline with the least environmental impact. This approach led to various treatment methods.
1. Manual Removal – Use of shovels, rakes, sorbents (removes oil by dabbing or wiping by hand over soiled area), and hand pickup to clean beach segments. Pooled oil, oily sediment and oily

Introduction

debris were recovered. This technique was used in lightly impacted areas and environmentally sensitive areas where using mechanical equipment is not desirable.

2. Cold Water Deluge – Pumping large volumes of ambient seawater to a perforated hose transporting loose oil downslope to the water line where it was trapped by booms and removed by skimmers or sorbents. Commonly used in heavily impacted areas in combination with warm or cold water wash.

3. Cold Water Wash – Pumping ambient sea water at high pressure through fire hoses, flushing oil down slope to be recovered by skimmers. Two common uses:

1) treat environmentally sensitive areas.
2) to congeal and transport oil mobilized during warm water washing.

4. Warm Water Wash – Up to 140 degrees F, high pressure seawater to mobilize weathered oil adhering to rocks. Cold water wash or deluge was typically used to congeal and float the oil to waterline for pickup. Used in areas where cold water wash and manual pickup were not effective in removing oil. This technique became the primary treatment for nearly all treated shoreline in Prince William Sound.

5. Bioremediation – Application of nutrients (fertilizer containing nitrogen and phosphorous) on the shore line to accelerate natural bioremediation of oil. Biodegradation is the natural process by which microorganisms attack petroleum hydrocarbons and reduce them to CO_2 and H_2O, and bioremediation is simply accelerating the process. This process was experimental and later proved to be beneficial in the speedy recovery of Prince William Sound and the Gulf of Alaska.

6. Natural Recovery – Leaving the beach untreated with removal of oil occurring by natural wave action and biodegradation. Used when alternative to "do nothing" policy would result in more harm than good.

7. Subsurface Injection – Used to remove subsurface oil which had penetrated into the sediment. Injectors deliver warm water or air into the substrate during high tide so oil would float or loosen to the surface for pickup.

8. Vacuum – Used in April when oil was mobile worked well when picking up pooled or gross oil deposits. Bulky equipment made vacuum procedure difficult to use for shoreline treatment.

9. Organic Sorbients – A special absorbent peat was used as an organic sorbent to soak up oil on impacted shoreline. Oil would strongly bind to the peat resulting in little or no re-oiling of rock surface. Peat also assisted in removal of oil in pinneped areas making rocks non-slippery.

Questions and Concerns

When you read this article consider the following:

Who do you think this article was written for? Was the author targeting the audience with ulterior motives perhaps.

Reading 1

The *Exxon Valdez* oil spill: Initial environmental impact assessment

Part 2 of a five-part series

Alan W. Maki
Exxon Company, U.S.A.
Anchorage, AK 99519-0409

The March 24, 1989, grounding of the *Exxon Valdez* on Bligh Reef in Prince William Sound, Alaska, was unprecedented in scale. So too was Exxon's response to the oil spill and the subsequent shoreline cleaning program, including the employment of more than 11,000 people, utilization of essentially the entire world supply of containment booms and skimmers, and an expenditure of more than two billion dollars. In the days immediately following the *Valdez* spill, Exxon mobilized a massive environmental assessment program. A large field and laboratory staff of experienced environmental professionals and internationally recognized experts was assembled that included intertidal ecologists, fishery biologists, marine and hydrocarbon chemists. This field program to measure spill impacts and recovery rates was initiated with the cooperation of state and federal agencies. Although the agencies subsequently foreclosed cooperation in most of these studies because of litigation concerns, this comprehensive assessment program continues today. Through the end of 1989, this program has resulted in well over 45,000 separate samples of water, sediment, and biota used to assess spill impacts.

It is the intent of this paper to provide initial observations and preliminary conclusions from several of the 1989 studies. These conclusions are based on factual, scientific data from studies designed to objectively measure the extent of the impacts from the spill. Data from these studies indicate that wildlife and habitats are recovering from the impacts of the spill and that commercial catches of herring and salmon in Prince William Sound are at record high levels. Ecosystem recovery from spill impacts is due to the combined efforts of the cleanup program as well as natural physical, chemical, and biological processes. From all indications this recovery process can be expected to continue.

Design of the assessment program

In early 1990 *Environmental Science & Technology* completed a four-part series on the state of the science of ecotoxicology and ecological risk assessment *(1–4)*. This series reviewed the development and current regulatory applications of the ecological risk assessment paradigm based on sequential comparisons of environmental fate (chemistry) data and biological effects data. Much of this literature is the basis for the Technical Information Documents *(5, 6)* being used by the regulatory agencies and supports the "Type B" natural resource damage assessment rules promulgated under Section 301 (C) of the Comprehensive Environmental Response, Compensation, and Liability Act (CERCLA).

The environmental fate-and-effects approach also forms the backbone of the *Exxon Valdez* environmental impact assessment program as discussed in these pages. Each of the studies was carefully designed to accurately measure hydrocarbon concentrations in the environment and within specific biological components as well as measure observed biological effects at individual trophic levels. For each program, environmental fate data complemented effects data to ensure that a fate-and-effects correlation could be developed to assess the likelihood of impacts.

Shoreline habitat

As a result of the grounding of the *Exxon Valdez*, some 258,000 barrels of crude oil were released into the water. Winds of more than 70 mph on the third day after the grounding rendered containment of oil on the water impossible, with the result that shoreline habitat in the southwestern segment of Prince William Sound was impacted. During the next few weeks more than 1100 miles of shoreline in south-central Alaska were impacted by oil to varying degrees (Table 1). These shoreline areas represented the habitat most obviously impacted by the spill where floating oil and mousse (oil-water emulsion) were washed ashore and deposited in intertidal areas. Indeed, it was these areas that were the focus of the two-billion-dollar cleanup program during 1989 and 1990.

It is important to point out that, at the very worst, slightly more than 10% of the shoreline habitat in Prince William Sound and the Gulf of Alaska was impacted by oil, thus leaving 85–90% of the regional shoreline untouched by oil (Table 1). Predictably, the percentage of impacted areas dropped significantly to

TABLE 1
Alaskan coast impacted by *Exxon Valdez* oil

Area	Total coast (miles)	Oil-impacted coast, May 1989 (miles)	Total coast impacted, May 1989 (%)	July 1990 oil-impacted coast (miles)[a]	Total coast with oil, July 1990 (%)
Prince William Sound	3061	357	11.6	91.5	3.0
Gulf of Alaska	6345	732	11.5	24.1	0.4
Total (miles)	**9406**	**1089**		**115.6**	

[a] Joint Exxon/Agency Spring Shoreline Assessment Teams (SSAT) report. Wide (>6 m), moderate (3-6m), and narrow (<3 m) oiling widths combined.

approximately 1% of the shoreline areas in 1990 as a result of both the cleaning program and natural weathering.

Shoreline oiling data are the results of the 1990 spring shoreline assessment wherein 20 separate teams, each consisting of state of Alaska, federal agency, and Exxon representatives, surveyed more than 1200 miles of shoreline in Prince William Sound and the Gulf of Alaska. A total of 822 miles of shoreline assessed had no oil; 115.6 miles contained oiling in <3–6 m or wider bands. In addition, 5071 pits were dug to determine the extent of subsurface oiling. Subsurface oil (oil deeper than 10 cm) was found in 733 (14%) of the pits dug. The locations recommended for 1990 summer cleaning were widely scattered and generally confined to short lengths of shoreline; all identified areas received attention during the summer.

Biota. The shorelines of southern Alaska are predominantly large rocks, boulders, and cobbles with less than 10% of the area characterized as soft substrate. This rocky intertidal zone is an extremely harsh habitat. Plants and animals living there have evolved to cope with a highly unstable substrate, extremely forceful breaking waves, and wide seasonal and diurnal ranges in photoperiod and temperature. As a result, many of these areas have naturally low species density and diversity.

Barnacles, mussels, and rockweed (*Fucus*) are the best indicator species for visually estimating the biological condition of these intertidal habitats. They are good indicators of trends in recruitment and loss caused by oil or other natural phenomena. In order to assess spill impacts on shorelines, these species and the intertidal communities in Prince William Sound and the Gulf of Alaska have been sampled via counts of species density, percent cover, species diversity, and photographic surveys of intertidal biota.

Results from the shoreline surveillance program indicate that although species densities were lower on heavily oiled shorelines, and high local mortalities occurred where oil coverage exceeded 90%, the species compositions of communities are similar to those that would be expected on comparable shorelines in the absence of oil. Even where oiling remains in sheltered, soft substrate areas, the spring and summer growing season of 1990 confirmed that continued recovery was under way. Thus, intertidal invertebrates and plants do survive on oiled shorelines, a point that supported spill researchers' conclusions that further extensive, intrusive shoreline cleaning efforts were unwarranted *(7)*.

This recruitment trend also has been observed in several studies of other spills worldwide that showed that shoreline biota recovered in impacted areas within a few years after the spill *(8)*.

Shoreline field-monitoring data support the following conclusions:

- The amount of oil remaining on the shoreline has continued to decrease since the summer of 1989. There are few oiled shores remaining.
- The major impact to intertidal communities was a decrease in the numbers of individuals of each species and not a decrease in the species composition of the communities. The intertidal communities are intact.
- Intertidal plants and animals were found settling and surviving on shorelines throughout Prince William Sound and the Gulf of Alaska throughout the 1989 and 1990 growing seasons.
- Even where residual oil remains on a few shorelines, biological recovery is taking place.

Water quality

Within six days of the spill, Exxon initiated an extensive water quality sampling program in Prince William Sound to help assess possible impacts on the marine biological community by measuring hydrocarbon concentrations in the waters of the Sound. Water samples were collected from March through October 1989 at 35 offshore locations.

More than 2300 water quality samples have been analyzed to date. Figure 3 is a plot of the average polycyclic aromatic hydrocarbon (PAH) concentrations throughout the year. The three curves represent the averages of different sites. The solid line represents the three sites with the highest concentrations. These are three bays that were heavily oiled. Note that the average of these three worst bays peaked at only 7% of the Alaska state water quality standard. The average of all primary sites has a maximum value of about 1/50th of the state allowable limit for aromatics in water *(9)*.

These data all confirm that average hydrocarbon concentrations measured in the water column for the more toxic components have consistently been well below state of Alaska standards and are 10–1000 times lower than levels lethal to plants and animals living in the water column, including commercially important fish species.

Fisheries

Pacific herring. Annually, millions of herring enter Prince William Sound to spawn in early April. This spawning period determines the commercial herring fishing season which had a 1988 value of $12 million. The *Valdez* spill occurred approximately three weeks prior to the peak of the Pacific herring spawn. Concern regarding potential impacts to the herring resulted from the fact that herring deposit their eggs on kelp in the intertidal and shallow subtidal zones in Prince William Sound.

Results of aerial surveys flown by Alaska Department of Fish and Game field biologists confirm that both 1989 and 1990 spawning activity was comparable to historical averages for the years immediately preceding the spill (Figure 4). These data provide convincing evidence that herring spawning activity has been neither impaired nor delayed during the period when highest potential existed for exposure to spilled oil.

These conclusions were further substantiated by the results of the 1990 commercial herring fishing season wherein 8300 tons of herring were harvested during the 20-minute season opening on April 12th—an all-time record catch for the sac roe fishery. Record harvests were also achieved in the gill net and wild roe-on-kelp fishery seasons, thus minimizing concerns over long-term impacts on this important commercial fish species.

FIGURE 3
Average polycyclic aromatic hydrocarbon concentrations for Prince William Sound water[a]

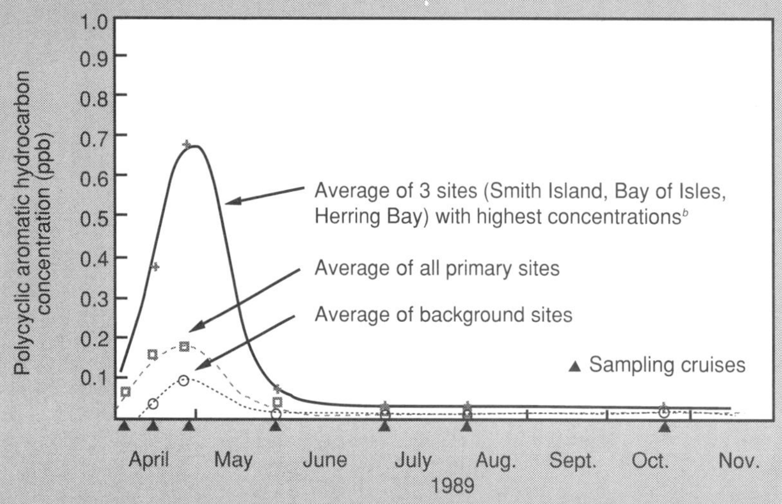

[a]State of Alaska criterion for petroleum aromatics in water is 10ppb.
[b]Peak is 7% of state allowable limit.

Pink salmon. The Prince William Sound salmon harvest is dominated by pink salmon and was worth approximately $63 million in 1988. Data collected during the critical life stages of the pink salmon population show no significant oil-related effects.

An extremely healthy and vigorous zooplankton bloom supported the spring 1989 release of more than 600 million fry from the commercial fish hatcheries of Prince William Sound. Outmigrating fry and juveniles in 1989 were apparently not impacted, and many juvenile salmon were observed in both oiled and unoiled nearshore areas. Similarly, field experiments with caged juveniles noted no differences in survival between oiled and unoiled sites. When the 1989 adults returned to spawn in the streams flowing into Prince William Sound, several streams had abnormally large numbers of returning fish. Also, no differences were evident in the spawning utilization of streams in oiled areas versus unoiled areas. Subsequent experiments in the fall with the incubation of eggs buried in gravel at several oiled and unoiled sites demonstrated similar survival and hatching rates between these sites, thus minimizing concerns over oil effects on this sensitive life stage.

Pink salmon have a short two-year life cycle. In the summer of 1990, those young salmon that left their native streams in April 1989, just after the spill, returned to spawn. The reported harvest through September 12, 1990, was 35.1 million fish taken, with an additional 8.9 million taken from hatchery stocks. This makes the 1990 commercial catch an all-time record, ahead of the 1987 record level by about 14 million fish and 163% ahead of the pre-season forecast for cumulative harvest. This strong run consists of both hatchery and wild-stock production (Figure 4).

In summary, there have been no indications of any significant pink salmon kills or effects on spawning activity related to oil exposure.

Shellfish and subsistence fisheries. Shellfish and crustaceans constitute a much less significant but still important part of Alaskan commercial, subsistence, and recreational fisheries. Several species of clams, mussels, crabs, and shrimp occupy the spill-impacted areas. Following the spill, a cooperative program was initiated between the Alaska Department of Fish and Game, the National Oceanic and Atmospheric Administration (NOAA), Exxon, and the subsistence communities to examine subsistence food safety in the spill-impacted areas. As of July 1990, more than 1300 samples of fish and shellfish have been collected in seven sampling cycles representing 23 species of fish and shellfish from 13 subsistence resource areas. They are being tested for aromatic hydrocarbons, the toxic components of crude oil most likely to be assimilated by shellfish.

The results of the chemical analysis received from the NOAA Seattle laboratory are very encouraging (Figure 5). Contaminant levels in all the fish and most shellfish were extremely low compared to data for fish and shellfish from a control site at the village of Angoon, outside the spill-impacted area. Except for shellfish taken from Windy Bay and from Kodiak Harbor, all food resources sampled showed safe hydrocarbon levels.

A Toxicological Expert Committee consisting of representatives of the Food and Drug Administration, the National Institute of Environmental Health Sciences, NOAA, the National Marine Fisheries Service, the Alaska Department of Public Health, and various university and industry toxicologists reviewed the data

FIGURE 4
Prince William Sound fish production

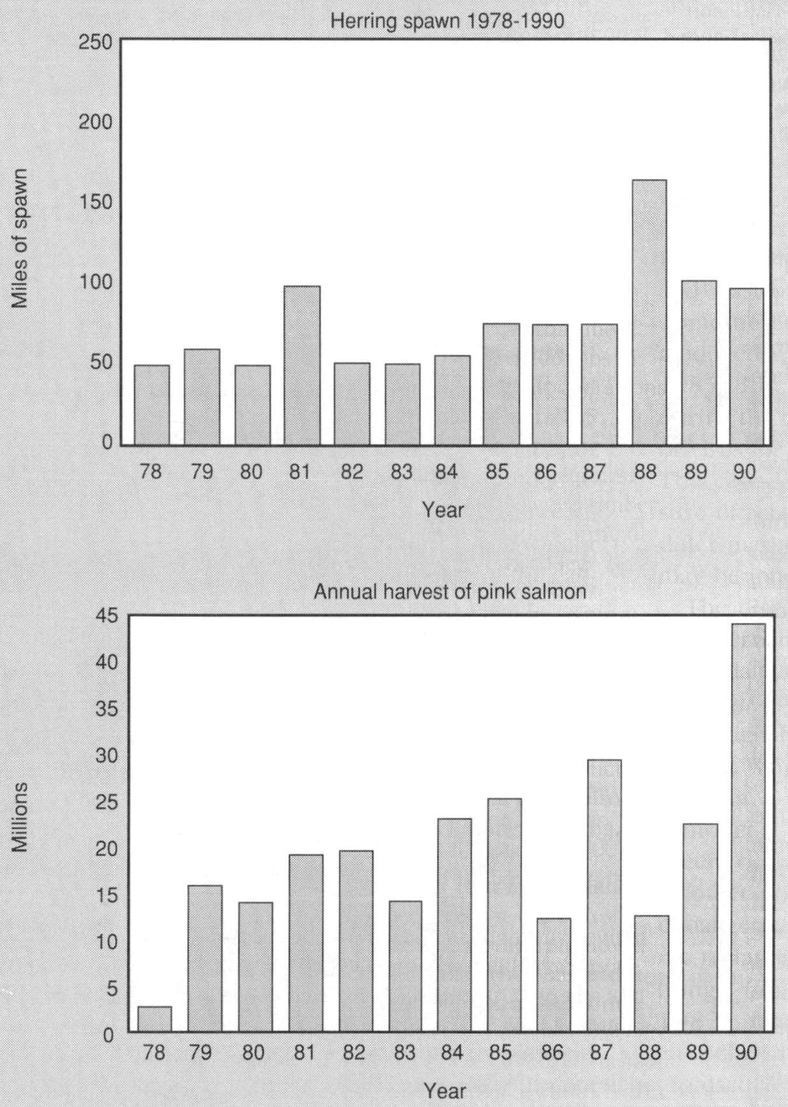

and filed a consensus report. The committee concluded that finfish from anywhere in the study area are safe to eat in unlimited quantities, and except for those taken from obviously oiled shores, shellfish are also safe to eat *(10)*.

Wildlife: Initial impacts

Some 36,000 bird mortalities were documented between March and September 1989. Although some of these were natural mortalities, no doubt exists that a significant number were caused by the spill. More than 1000 sea otter and 153 bald eagle mortalities were initially attributed to the spill. Although there was an extensive effort to recover dead birds and animals, actual mortalities were likely higher. When the above mortality figures are compared with total populations in the spill-impacted area—more than 10,000,000 sea birds; more than 30,000 sea otters; and 5000 eagles—indications are that recolonization has been rapid and robust, as it has been following other spills throughout the world *(8)*.

Furthermore, extraordinary measures were taken to rescue and care for birds and otters that were impacted. From the period of March through September 1989, Exxon organized the largest and most comprehensive bird and sea otter rescue and rehabilitation programs ever attempted. Facilities for the holding, cleaning, and care of oiled birds and otters were built at Valdez, Seward, Homer, Kodiak, and Anchorage. During the program, more than 140 boats and 5 aircraft were used to retrieve oiled birds and otters from remote locations throughout Prince William Sound and the Gulf of Alaska.

Some 71 species of birds were handled during the six-month program. More than 1600 birds were brought in live for treatment with a release rate for the sea birds of 50%. Given the time of year of the spill, the climate, the remoteness of the site, and the logistics, this survival and release rate compares favorably with the 30–60% range of release rates seen in other spills. At its maximum, the bird program employed some 400 bird rescue personnel; this effort cost more than $25 million.

Teams were hired to work in Prince William Sound and Kodiak to collect eagles from heavily impacted areas to determine the need for potential rehabilitation. These capture teams caught 114 eagles; only 16 required treatment. (An additional 23 eagles caught by others not associated with the capture teams were also treated).

The staff of the otter centers grew to more than 320 specialists and volunteers at all three locations. From March 30 until September 14, 357 sea otters were treated and held at the centers; 223 (62%) were rehabilitated and released or placed in aquariums. Expenditures for this effort exceeded $18 million.

Wildlife monitoring program

As noted, in spite of the best efforts by all parties, many birds died. To assess overall population-level impacts, comprehensive wildlife monitoring programs were initiated after the spill and are ongoing.

Bird monitoring. Results of Exxon and agency winter, spring, and summer wildlife monitoring programs are very

FIGURE 5
Concentrations of aromatic hydrocarbons[a]

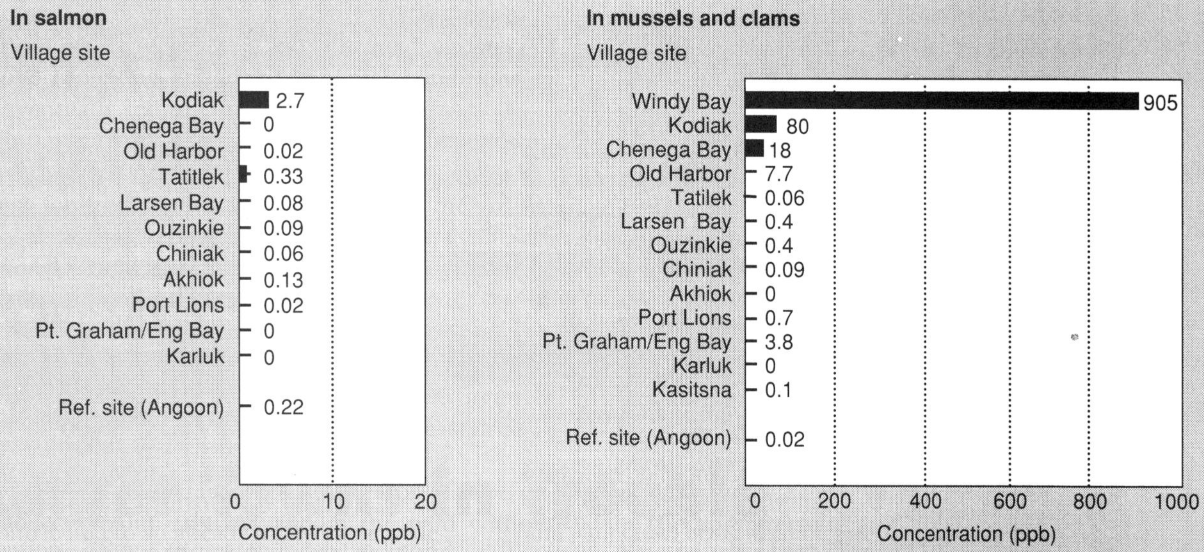

[a]Prince William Sound, lower Cook Inlet, Kodiak.

encouraging. There have been no documented reports of oil or sheens affecting populations of birds or marine mammals, and there have been no confirmed mortalities attributable to oil since September 1989. Additionally, wildlife observations in Prince William Sound indicate that significant numbers of birds from 45 different species occupy previously heavily oiled areas and appear to be unaffected by existing conditions. Species, diversity and density are similar for both oiled and unoiled areas.

The findings of the winter wildlife population surveys are most encouraging. These surveys of oiled areas in Prince William Sound show that bald eagle sightings have consistently increased since the summer of 1989. Data from cooperative U.S. Fish and Wildlife Service eagle surveys throughout the summer of 1990 indicate that 1031 of 2030 nests (51%) in the spill area were active and that 61 of 75 previously active nests (81%) in Prince William Sound cleanup sites produced an average of 1.4 eaglets per nest, thus matching historical productivity statistics for this area. It is recognized these data do not allow definitive statements of eagle population dynamics, but they do provide conclusive evidence that animal life is present in habitats previously impacted by the oil spill.

Mammal monitoring. Results of the winter mammal monitoring program indicate the presence and successful reproduction of mink and river otters in both oiled and unoiled bays of Prince William Sound. Sea otters are present in all areas in apparently equal numbers in oiled and unoiled bays.

Although census results of sea otters often vary widely, the Prince William Sound population has been generally estimated at 5000 to 6000 *(11)*, with some recent estimates being as high as 9000. The Prince William Sound population was believed to be less than 50 inidividuals when commercial harvesting of this species ended in 1911. The population has increased very rapidly in recent years, and field surveys of 550 miles of oiled shorelines counted more than 350 sea otters. This survey covered about one-sixth of the total Prince William Sound shoreline. It indicates that otter populations are approaching pre-spill densities *(11)* and provides convincing evidence of sea otter recolonization of previously heavily oiled areas.

In summary, the monitoring data for mammals indicate that environmental conditions have improved enough that the spill should have no further substantive impacts on wildlife. Recovery is well under way.

Conclusions

A recently published Congressional Research Service report *(12)* carefully reviews the extent of environmental impacts from several previous oil spills worldwide and concludes that "To date, pollution from offshore petroleum activities has not appeared to be a significant threat to the survival of various species . . . Despite short-term media attention to the catastrophic nature of major spill events, the chemicals contained in petroleum have long been part of the marine environment and physical impacts are likely to be temporary in the dynamic natural flux of the coastal environment."

Available data to date for the *Valdez* spill are consistent with the observations made in this report. Samplings of petroleum aromatic hydrocarbon concentrations in the waters of Prince William Sound clearly demonstrate that average levels have remained well below exposure levels known to cause acute and chronic effects to sensitive aquatic life.

Field counts of plants, fish, and mammals from throughout the spill area provide convincing data that wildlife species are surviving and reproducing, thus confirming that biological recovery is rapidly taking place.

References

(1) Bascietto, J. et al. *Environ. Sci. Technol.* **1990,** *1,* 10–14.

(2) Cairns, J.; Mount, D. *Environ. Sci. Technol.* **1990,** *2,* 154–61.

(3) Hoffman, D.; Rattner, B.; Hall, R. *Environ. Sci. Technol.* **1990,** *3,* 276–83,

(4) Harris, H. et. al. *Environ. Sci. Technol.* **1990,** *5,* 598–603

(5) "Injury to Fish and Wildlife Species"; Type B Technical Information Document, CERCLA 301 Project; U.S. Department of the Interior: Washington, DC, June 1987.

(6) "Guidance on Use of Habitat Evaluation Procedures and Suitability Index Models for CERCLA Application"; Type B Technical Information Document, CERCLA 301 Project; U.S. Department of the Interior: Washington, DC, June 1987.

(7) "Excavation and Rock Washing Treatment Technology—Net Environmental Benefit Analysis"; compiled by Hazardous Materials Response Branch, National Oceanic and Atmospheric Administration: Seattle, WA, July 1990.

(8) Baker, J.; Clark, R.; Kingston, P. *Natural Recovery of Cold Water Marine Environments After an Oil Spill: Environmental Recovery in Prince William Sound and the Gulf of Alaska;* 13th annual Arctic and Marine Oil Spill Program; Edmonton, AB, Canada, June 1990.

(9) Neff, J. "Water Quality in Prince William Sound"; Special Report of Battelle Ocean Sciences Laboratory: Duxbury, MA, April 1990.

(10) "Summary of Findings of Toxicological Expert Committee for Evaluating Data Related to the Consumption of Marine Subsistence Foods (*Exxon Valdez* Oil Spill)"; National Oceanic and Atmospheric, Administration: Seattle, WA, Feb. 21–22, 1990.

(11) Irons, D.; Nysewander, D.; Trapp, J. *Prince William Sound Sea Otter Distribution in Relation to Population Growth and Habitat Type;* Alaska Investigations Field Office. U.S. Fish and Wildlife Service; Anchorage, AK, April 1988.

(12) Mielke, J. *Oil in the Ocean: The Short and Long-Term Impacts of a Spill;* Congressional Research Service; Library of Congress: Washington, DC, July 24, 1990.

Alan W. Maki *is currently senior science advisor to Exxon Company, U.S.A., in Anchorage, AK. He received a B.S. degree from the University of Massachusetts, an M.S. degree from the University of North Texas, and his Ph.D. in fisheries and wildlife management from Michigan State University. His research interests have focused on the development of applied ecotoxicology and ecological risk assessment. He has served on numerous EPA and National Academy of Sciences advisory panels and is a former president of the Society of Environmental Toxicology and Chemistry. He has been stationed in Alaska for the past five years.*

Discussion for The Exxon Valdez Oil Spill

1. Do you think the recovery time would have been shorter if the spill had been left alone?

2. If you were the President of Exxon would you have done anything different?

3. If you visited the site today do you think you would see any evidence of the damage?

4. Do you think the Exxon company has a lasting responsibility to the area?

5. No evidence was found indicating that the whales were impacted by the spill. Design a research program to test this.

6. What safety rules should be instituted for this area in the future?

7. Birds were heavily impacted in this spill. Can you design an emergency response team for the future? What equipment, staff and supplies would you take?

8. An oil spill here in Alaska is different from a spill in the Middle East. Can you discuss why?

Introduction

Hazardous Waste: Not in Our Backyard!

This exercise examines the impending grid lock that is about to fall upon the researchers, medical laboratories, and energy producers of California. Your well being may depend upon the outcome of this environmental (or is it hysterical?) confrontation.

Barry Thomas

In January 1993 a San Francisco Federal Judge stopped the transfer of land near Needles California from the Federal Government to the California State Government. This stopped, for the time being, the development of California's first and long awaited low level nuclear-waste dump. This was a win for the anti-dump proponents and a set back for industry, hospitals and research institutions throughout the Northwest. Anti-dump environmental groups cited the need to protect the endangered desert tortoise and development of the 1000 acres would adversely impact the tortoise habitat. The underground aquifer was another concern since pollution of this would result in contamination of the Colorado River. This action throws the land swap decision into the hands of the new Interior Secretary President Clinton appointed, Bruce Babbitt.

The Beatty, Nevada and Richland, Washington facilities stopped taking Californian waste December 31, 1992. At this time low level nuclear waste generated in California is handled through the Southeast Compact Commission and the Barnwell disposal facility in South Carolina. Agreements are in place to handle West Coast waste through June 30 1994. All out of state generators of low level radio-active waste will be charged an access fee of $220 per cubic foot for the privilege of dumping plus disposal charges levied by the site operator and the transportation charges. For large generators the contract will be for a set volume and the payment is to be in advance, payable in six quarterly payments. Any waste in excess of this prearranged volume will be charged at 130-150% ($286-$330 plus fees per cubic foot). The total fee will be payable regardless of the volume shipped. Such costs are passed on to the generator and the users of those facilities. New York is expecting charges of $1,100 per cubic foot for their low level radio-active waste.

Such costs will cripple biomedical research in that state since 30-40% of low level radio-active waste is from hospitals. Here in the Northwest medical services and educational facilities will also be hard pressed to cover such costs. The lack of a suitable dump site has made storage a serious problem. Nuclear power plants ship 8-10,000 cubic feet of radio-active waste each year and hospitals and universities each ship 10 or more barrels a year. California alone is producing about 300,000 cubic feet of low level radio-active waste each year. Today, a rat contaminated with a radio-active tracer costs nearly double its purchase price to dispose of, yet if that amount of radio-active material was in solution then it could probably be legally flushed down the drain. Gloves and any suspected contaminated supplies are required to be treated as LLRA waste. Probably greater than 80% of what is being dumped is uncontaminated waste.

One other major headache is what to do about waste that was dumped years ago. How do we handle it today? If it is a potential disaster waiting to happen who has the responsibility and the authorization to handle the problem? If there is no available dump site then what will happen and who is to blame when the disaster hits? The last reading will elucidate such a potential disaster.

Our regulations have contributed to hysteria and misinformation about the dangers of this material. Fears of the general public make finding a solution to this problem difficult. Switching to short life agents and minimizing waste has not abrogated the need for a safe disposal site. What is needed is proof of a safe system and an education program that can instill a sense of security in the general public, particularly those who live next to these proposed dump sites. The following readings will elucidate the generalized fears of the public and the responses to those fears from professionals.

Questions and Concerns

READING 1

As you read this article consider the following:

1. Is this a comprehensive analysis of the situation?
2. Has the State of California fulfilled its obligations?
3. What can be done to get things moving?

Reading 1

The original version of this Practical Guide was published in the California Environmental Insider (May 15, 1992).

AN UPDATED PRACTICAL GUIDE TO THE MANAGEMENT OF LOW-LEVEL RADIOACTIVE WASTE IN CALIFORNIA AFTER JANUARY 1993

by Jennifer L. Hernandez, Esq. and Alan D. Pasternak, Ph.D.

In January of 1993, California's research universities, biotechnology companies and other generators of low-level radioactive waste (LLRW) lost assured access to existing disposal facilities in Washington, Nevada and South Carolina. California's efforts to site its own disposal facility have been delayed pending resolution of state and federal lawsuits. California generators must now pursue one of three options: (1) interim storage of LLRW in their own facilities; (2) interim storage at one of the scarce commercial storage facilities; or (3) until July 1, 1994, very costly and uncertain access to the South Carolina LLRW disposal facility. Many of the affected LLRW generators have attested that loss of assured access to LLRW disposal facilities will increase the cost -- and ultimately impair California's ability to develop and administer -- existing and potential new treatments for AIDS, cancer, and other diseases. This Practical Guide provides a brief summary of the background and current status of California's efforts to fulfill its legal obligation to provide for adequate LLRW disposal capacity, and provides generators with practical recommendations for handling their LLRW after January 1993.

1. What Is Low-Level Radioactive Waste (LLRW)?

LLRW is essentially trash contaminated with radioactivity, such as clothing, tools, laboratory equipment like glassware and filters, and parts of machines or devices that use radioactive materials. LLRW does not include uranium mill tailings or high level waste such as fuel rods from nuclear power plants. The majority of LLRW is generated by biotechnology companies, universities, hospitals and clinics during research activities, with the remainder generated by government agencies and utilities. (Non-radioactive chemical "hazardous wastes" are regulated under the California Hazardous Waste Control Laws and the Federal Resource, Conservation and Recovery Act.) LLRW is separated into Classes A, B and C, with hazardous lives of 100, 300 and 500 years, respectively. Radioactive waste which is "greater than Class C waste" is not LLRW.

2. How Is LLRW Regulated?

A. **Regulation of Radioactive Materials Generally.** The Atomic Energy Act of 1954 (42 USC § 2014, *et seq.*; "AEA") establishes a comprehensive regulatory program for the use and disposal of radioactive materials. LLRW is a subset of the materials regulated under the AEA. In California, most users of radioactive materials which generate LLRW must receive a license from the Department of Health Services ("DHS"). Pursuant to a delegation agreement with the Nuclear Regulatory Commission ("NRC"), DHS administers the AEA's user license and equipment registration program for most of California's radioactive materials users other than the Department of Defense ("DOD"), the Department of Energy ("DOE"), and the nuclear power plants. Detailed safety, handling, training and recordkeeping requirements for radioactive materials are set forth in the NRC regulations

GRAHAM & JAMES
Environmental
Practice Group

San Francisco
One Maritime Plaza
Third Floor
San Francisco, CA 94111
(415) 954-0200

Palo Alto
525 University Avenue
Fourteenth Floor
Palo Alto, CA 94301-1977
(415) 327-0500

Los Angeles
725 South Figueroa Street
Thirty-Fourth Floor
Los Angeles, CA 90017-5434
(213) 624-2500

Sacramento
555 Capitol Mall
15th Floor
Sacramento, CA 95814
(916) 552-6600

(*see generally*, 10 CFR Part 20) and corresponding DHS regulations (*see generally*, 17 CCR § 30100, *et seq.*).

B. **Regulation of LLRW.** In the Low Level Radioactive Waste Policy Act of 1980 and the Low-Level Radioactive Waste Amendments Act of 1985 (collectively referred to herein as the "Act") (42 USC § 2021b, *et seq.*), Congress adopted a comprehensive system designed to ensure both regional equity in disposal and adequate nationwide LLRW disposal capacity. (*See generally*, 10 CFR Part 61) There are two critical concepts in the Act:

1. **Each State is Responsible for Developing Adequate LLRW Disposal Capacity.** Each state is responsible for ensuring adequate disposal capacity for LLRW generated within the state. (Act § 2021c(a)(1)(A). States are also encouraged to enter into regional, multi-state compacts to encourage the development of cost-effective regional LLRW disposal facilities. (Act § 2021d(a)(2).)

2. **States Which Fail to Provide Adequate LLRW Disposal Capacity Are Subject to Certain Sanctions.** The Act includes specific milestones for measuring each state's progress in developing LLRW disposal capacity and includes sanctions for states which fail to meet these milestones. (Act § 2021(e)(e).) The Act includes three forms of sanctions:

   a. **Loss of Access to LLRW Disposal Facilities.** To ensure that all states and compacts would develop adequate disposal capacity, and to encourage states to enter into regional LLRW disposal compacts, Congress explicitly authorized compacts to restrict access to their LLRW disposal facilities to the member states of the compact. (Act § 2021e(a)(3).) Generators in states or compacts which failed to meet various milestones could be denied access to LLRW disposal facilities in other compacts. (Act § 2021d(c).

   b. **Monetary Incentives.** States or compacts with an existing LLRW disposal facility ("sited states") were authorized to charge specified disposal surcharges to generators from non-sited states. (Act § 2021e(d)(1)). A portion of these surcharges were forwarded to DOE and placed in an escrow account to assist non-sited states in developing LLRW disposal capacity. (Act § 2021e(d)(2)(A)). States or compacts which failed to meet specified milestones for providing LLRW disposal capacity forfeit their right to reimbursement of these LLRW disposal surcharges. Early milestones could be met by making progress in planning to provide LLRW disposal capacity; the final milestone for this monetary incentive provision is January 1, 1993, and can be met only by states or compacts which actually provide generators with LLRW disposal capacity. States and compacts failing to meet this final milestone are required to forfeit these surcharges, which are then to be paid out to the LLRW generators in these states. (Act § 2021e(e)).

   c. **Take Title Incentive.** The Act's most significant incentive required that states which failed to provide LLRW disposal capacity by January 1, 1996 were required to "take title" to the LLRW generated in that state. (Act § 2021e(d)(2)(C)).

3. **Take Title Provision Held Unconstitutional.** New York challenged the constitutionality of the Act's incentive structure as a coercive effort to "commandee[r] the legislative processes of the States by directly compelling them to enact and enforce a federal regulatory program." (*New York v. United States* 942 F.2d 114 (2nd Cir. 1991), *aff'd in*

part, rev'd in part, 505 U.S. ___, 112 S.Ct. 2408 (June 19, 1992); citing *Hodel v. Virginia Surface Mining & Reclamation Assn., Inc.*, 452 U.S. 264, 288 (1981).) The district and appellate courts ruled against New York; the Supreme Court earlier this year reached a mixed ruling. In a majority opinion by Justice O'Connor, the Court held that the access and monetary incentives were appropriate exercises of the federal government's Commerce and Spending clause authority. The "take title" provision, in contrast, gave states a "choice" only between two unconstitutionally coercive alternatives -- regulating LLRW according to the rules established by the federal government *or* taking title to the LLRW -- and was thus struck down. The Court's decision left LLRW generators with the worst of both worlds: they could be denied access to existing LLRW disposal facilities but could not force the state to take title to the LLRW.

3. What Has California Done To Satisfy Its Legal Obligations Under The Act?

California was well on its way to meeting its legal obligations under the Act, and had declared in December 1991 that it was ready to make the final decisions necessary to open a LLRW facility to serve California generators and generators from the other three states in the Southwestern Compact. This process was hijacked when DHS was forced to agree to conduct a lengthy adjudicatory hearing in order to secure Russell Gould's Senate confirmation as Secretary for the Health and Welfare Agency and Molly Coye's confirmation ad Director of the Department of Health Services. A brief overview of California's ongoing LLRW saga follows:

A. Enactment of Senate Bill 342 in 1983. This bipartisan urgency legislation (carried by Alquist, D/San Jose and Bergeson, R/Newport Beach) required DHS to develop a LLRW disposal facility to be operated by a private contractor on land owned by the state or federal government. DHS was also ordered to invite other states to join California in forming a regional disposal compact.

B. Creation of Southwestern Compact in 1987. Arizona, North and South Dakota joined California in forming the "Southwestern Compact," with California assuming responsibility for developing the first regional LLRW disposal facility. This facility is to provide thirty years' disposal capacity for the use of Southwestern Compact state generators. (*See generally*, Southwestern Low-Level Radioactive Waste Disposal Compact Consent Act, Pub. L. 100-712, and Cal. Health & Safety Code § 25877, *et seq.* (collectively referred to herein as the "Compact Act").)

C. Siting Proposed LLRW Disposal Facility. In 1985, following a competitive bidding process, a waste disposal company called US Ecology ("USE") was selected by DHS to be the "license designee" charged with developing, operating and eventually closing the LLRW disposal facility under the Compact Act. The first step in this process was to identify a suitable location for the facility. Implementing siting criteria developed by DHS, USE worked with a Citizens Advisory Committee which included representatives of the League of Women Voters and the Sierra Club, various regulatory agencies, and other interested parties including representatives of the Board of Supervisors of San Bernadino, Inyo and Riverside Counties. Following detailed studies and numerous public hearings, the "Ward Valley" site was identified as the preferred site. This site is located on federal lands in a remote area designated as a utility corridor under the California Desert Conservation Plan. The site is administered by the Federal Bureau of Lands Management ("BLM") and is located approximately 20 miles from Needles in San Bernardino County.

D. Completion of License Application and Environmental Analysis. DHS and BLM completed a joint Environmental Impact Report ("EIR")/Environmental Impact Statement ("EIS") on the Ward Valley project (and several alternative sites) in April of 1991. The EIR/EIS addressed siting, construction, operational, closure and environmental mitigation issues. The EIR/EIS concluded that all potentially significant project impacts would be mitigated to a less than significant level. Later in 1991, DHS conducted another series of public hearings on the pending USE license application to operate Ward Valley.

E. **DHS Progress on Ward Valley Delayed By Senate Rules Committee.** In December of 1991, DHS confirmed in writing that it had received all necessary documentation, and completed all required procedures, to make a final decision on the certification of the Ward Valley EIR and the issuance of the Ward Valley license. In an extraordinary series of sessions before the Senate Rules Committee ("Committee"), the Governor's appointee as Secretary of the Health and Welfare Agency, Russell Gould, first refused, but eventually agreed, to require DHS to hold an adjudicatory hearing prior to allowing DHS to make a final decision on the Ward Valley license. Dr. Molly Coye was forced to make a similar pledge as a condition to being confirmed to serve as Director of DHS. The transcript of the Committee proceedings leaves no doubt that Mr. Gould and Dr. Coye would not have been recommended for confirmation by the Committee if they had not promised to hold the adjudicatory hearing. While Mr. Gould and Dr. Coye pledged to limit both the scope and time required for such a hearing, neither the hearing procedure nor the hearing schedule have been established to date. The Committee has asserted an ongoing right to negotiate with DHS regarding the appropriate hearing procedures, seeking (among other procedures which deviate from the ordinary adjudicatory procedures established under California's Administrative Procedures Act (Cal. Gov't Code §§ 11500, *et seq.*, "Cal/APA") to require DHS to hire an "independent" judge acceptable to the Committee rather than use one of the DHS administrative law judges.

F. **Lawsuit Filed Challenging Coerced "Agreement" To Require Adjudicatory Hearing.** In July 1992, the California Radioactive Materials Management Forum, USE, and the National Cancer Patients, each filed Petitions for a Writ of Mandate and Prohibition with the California Supreme Court, asking the Court to take original jurisdiction, to immediately stay the adjudicatory hearing on the Ward Valley license, and to order DHS to make its license decision without a further hearing. The Supreme Court referred the matter to the Third Appellate District, which consolidated the three actions, granted the Committee's request to intervene as a Real Party in Interest ("RPI"), and held a hearing on September 15, 1992. At the hearing, the parties addressed the following issues:

1. Did the Committee violate the separation of powers and presentment clauses of the Federal and State Constitutions by conditioning Mr. Gould's and Dr. Coye's confirmation on their agreement to hold an adjudicatory hearing on the Ward Valley license?

    - Petitioners argued that the "agreement" to hold the hearing was unlawfully coerced in violation of the Constitutions.

    - Respondents (Mr. Gould, Dr. Coye, DHS and the Health & Welfare Agency) were represented by the California Attorney General's office. The Attorney General agreed that DHS had not previously held an adjudicatory hearing on a radioactive materials license application, that to the extent DHS had the discretion to hold such a hearing in the Ward Valley license proceeding it had previously exercised this discretion by not holding a hearing (and had already successfully defended this decision in an earlier lawsuit), and that a stay of the adjudicatory hearing was appropriate pending the Court's final resolution of the Petitions.

- The Committee argued that the Committee's confirmation authority was properly exercised both in conditioning the confirmations on an agreement to hold an adjudicatory hearing and in ongoing Committee involvement in establishing appropriate hearing procedures.

2. Was an adjudicatory hearing required by law and, if not, did DHS have the discretion to require an adjudicatory hearing under these circumstances?

   - Petitioners argued that an adjudicatory hearing was not required by law for the issuance of radioactive materials license, noting that no such hearing had been required for the thousands of licenses issued in the nearly 30 years since the hearing requirement was enacted. (Cal. Health & Safety Code § 25845(a)). Petitioners also contrasted the hearing required for the issuance of licenses -- in which any party could require DHS to hold a hearing "on the record" -- and the hearing required for license revocations or suspensions -- in which a Cal/APA hearing was required -- as further evidence that the Legislature did not intend to require an adjudicatory hearing for the issuance of licenses. (Cal. Health & Safety Code § 25845(a) and (b), respectively).

   - Respondent agreed with Petitioners that the applicable hearing statute did not require an adjudicatory hearing but noted that DHS had the discretion to require such a hearing under other statutory authority.

   - The Committee argued that "on the record" constituted "magic words" which required an adjudicatory hearing under the Federal Administrative Procedures Act (5 U.S.C. § 551, *et seq.*). The Committee argued that the absence of any citation to the Cal/APA in the license issuance hearing statute meant that DHS could not take advantage of some of the features of the Cal/APA, such as the use of DHS's own administrative law judges.

Two days after the hearing, the Court issued an injunction to prevent commencement of the challenged adjudicatory hearing, denied all other pending motions for RPI status, and issued an Alternative Writ ordering DHS to grant the relief requested by the Petitions or show cause why such relief should not be granted. On October 6, the Attorney General filed a return which noted that it had nothing to add to its previous pleadings. Governor Wilson also filed a lengthy *amicus* brief which disagreed with the Committee's interpretion of the scope of its confirmation authority. All parties completed the required briefs by November. The matter is now pending before the Court.

G. **Land Transfer Controversy.** The Compact Act also provided that the LLRW disposal facility must be located on land owned by the state or federal government. (Cal. Health & Safety Code § 25812). Under the original plan for developing a LLRW disposal facility, California planned to utilize the "Indemnity Selection" mechanism for transferring the Ward Valley site to the state. (Cal. Pub. Res. §§ 6201, *et seq.*). "Indemnity Selection" is one of several available mechanisms for transferring federal lands to the state, but is unique in that only the State Lands Commission ("SLC") is authorized to initiate this form of land transfer. In the Spring of 1991, however, SLC informed BLM and other interested parties that it was no longer willing to participate in this process and that another land transfer mechanism should be utilized.

In the summer of 1992, DHS applied to BLM for a direct sale of the Ward Valley parcel to the state under the Federal Land Policy and Management Act (43 USC §§ 1701, *et seq.*). BLM accepted the DHS application and formally turned down the still-pending SLC indemnity selection request on the grounds that SLC had offered insufficiently valuable land in trade for the Ward Valley parcel.

Reversing its 1991 determination to have no further involvement in the Ward Valley land transfer, in September of 1992 the SLC staff filed a new Indemnity Selection application which added additional lands in trade for the Ward Valley site in an effort to correct the technical defect noted by BLM in its rejection of the original SLC Indemnity Selection application. In January 1993, Secretary Lujan responded to Governor Wilson's request to complete the land transfer prior to leaving office. On January 19, 1993, a federal judge prevented the land transfer from taking place prior to the change in administration in response to a lawsuit filed by project opponents under the federal Endangered Species Act. Another federal court lawsuit challenging the land transfer was also filed on January 1993 by Grey Davis on January 19, 1993.

H. **Potential Access to Southeast Compact LLRW Disposal Facility**. The LLRW disposal facility located in Barnwell, South Carolina, has been made conditionally available to California generators. After a heated political debate conducted in the shadow of a significant state budget deficit, the South Carolina legislature voted to allow Barnwell to continue to accept out-of-compact LLRW for an eighteen-month period beginning in January 1993 and ending in July 1994 under certain terms. The terms require California or the Southwest Compact to enter into an agreement with the Southeast Compact under which mid- to large-sized generators of LLRW are required to identify projected LLRW volumes to be sent to the Barnwell facility, and are required to pay for those volumes without regard to whether they actually send any LLRW to Barnwell. Generators must also file a petition to export LLRW to Barnwell with the Southwestern Compact Commission, must obtain the necessary licenses from South Carolina, and must enter into a contract with the commercial operator of the Barnwell facility.

4. What Specific Issues Remain Unresolved For The Ward Valley Facility?

Opponents of the Ward Valley project have cited a range of issues from minor technical questions to fundamental political objections to the Act and the Compact Act. These include:

- **Ward Valley Safety and Environmental Issues.** The adequacy of the NRC's regulations on LLRW packaging, site monitoring, and other safety and environmental concerns have been the subject of numerous hearings. The EIR/EIS for the Ward Valley facility found that all environmental impacts would be mitigated to a less than significant level if the project was built, operated and ultimately closed in conformance with specified plans, license conditions and mitigation measures. USE has been found qualified to develop and operate the facility as the state's license designee. In addition to the EIR/EIS and licensing procedures, in 1991 DHS convened a Blue Ribbon panel, consisting of experts in facility design and groundwater protection, which was drawn from governmental agencies, the Sierra Club, and other private and governmental groups from all over the country. This panel concluded that the measures to be utilized at the Ward Valley project were appropriate and satisfactory from a regulatory, safety and environmental perspective.

- **Local Agency Reimbursement.** On a local level, the City of Needles continues to question the methodology for calculating the size of a fund created by the Act for the purpose of reimbursing local governments for costs associated with the development and operation of a LLRW disposal facility. The City of Needles is also examining the potential for extracting the groundwater located approximately 650 feet beneath the Ward Valley site.

- **Liability of LLRW Generators.** Another concern regards the applicability of other environmental laws, particularly the generator liability provisions of the Comprehensive Environmental Response, Compensation and Liability Act (42 USC § 9601, *et seq.*), to the Ward Valley facility. The United States Environmental Protection Agency has confirmed that unlawful releases of radioactive substances from LLRW disposal facilities such as Ward Valley are covered by CERCLA's liability provisions.

- **Political Challenges to Act and Compact Act.** Even more fundamental political challenges have been raised regarding the Act and the Compact Act. Several bills were considered last year relating to the Ward Valley project but none were enacted.

5. What's The Bottom Line For California LLRW Generators?

Although the current status of several issues relating to the development of the Ward Valley site is relatively complicated to track and assess, as a practical matter California LLRW generators should implement alternate strategies for managing LLRW beginning in 1993. In particular, generators should consider the following sequence of issues and potential actions:

- Identify Scope of LLRW Issues. Evaluate current projects and processes to identify and quantify LLRW to be generated within the next two to five years. Determine the maximum amount of LLRW currently authorized for onsite storage. Compare this amount to anticipated future LLRW waste volumes, and identify any shortfall in LLRW storage capacity that may arise over a two to five year period.

- Evaluate Onsite Options. Determine whether applicable facility permits allow storage increased quantities of LLRW onsite. At minimum, review radioactive materials handling license, business plans and related local hazardous material storage ordinance permits, conditional use permits, and the applicable requirements of the California Environmental Quality Act. (Cal. Pub. Res. Code §§ 21000 *et. seq.*) Determine whether the facility lease, insurance policies, or other private party agreements allow storage of increased quantities of LLRW onsite. Identify the physical capacity and suitability of increasing the quantity of onsite LLRW storage, and identify the necessary permit(s) and authorization(s) that needed to store LLRW onsite for a two to five year period. Consider the costs, benefits, schedule and likelihood of successfully undertaking the permit application/lease amendment/construction processes which would be needed in order to store shorter half-life LLRW for decay, and to store longer half-life LLRW on an interim basis pending the availability of a safe, cost-effective LLRW disposal facility.

- Evaluate Commercial Interim Storage Facility Options. Identify available off-site, third party facilities authorized to store LLRW. (A shortage of these facilities has already been confirmed in California.) Evaluate the price of interim storage, conduct an independent evaluation of each candidate facility's licenses, and arrange for facility inspections to evaluate safety and environmental compliance issues. Consider the costs and benefits of entering into a contractual arrangement with such a facility to provide for interim handling of LLRW pending the availability of a safe, cost-effective LLRW disposal facility.

- Evaluate Costs and Risks of Temporarily Disposing of LLRW at the Barnwell Facility. Review terms of Southwestern Compact Petition, required South Carolina licenses, and transporter/operator contracts needed to gain temporary access to LLRW disposal facilit in

Barnwell. Evaluate the price of interim storage, conduct an independent evaluation of the Barnwell facility's licenses, and arrange for facility inspection by independent third party to evaluate safety and environmental compliance issues. Consider the costs and benefits of temporarily disposing of LLRW at the Barnwell facility.

- Compare Onsite, Interim Storage Facility, and Temporary Barnwell Options. Compare the costs (including liability risks) and benefits of each option. Select a preferred option, or pursue two or more options on a parallel track (eg., completing the necessary paperwork to gain temporary access to Barnwell but delaying shipments to Barnwell until other options become infeasible), to maximize the likelihood of having a safe and lawful means of handling LLRW in 1993 and beyond.

6. How Can California Generators Encourage The State To Move Forward With the Ward Valley Facility?

The California Radioactive Materials Management Forum includes representatives from colleges and universities, hospitals, biotechnology companies, utilities and other industries which generate LLRW. The Forum has been instrumental in representing the interests of the generators during all phases of California's efforts to comply with the Act. The Forum continues to work with the members of the state's executive and legislative branches to bring the lengthy, complex and expensive process of siting and licensing the Ward Valley facility to a successful conclusion. Generators interested in learning how they can become involved in this process should contact the Forum (510/283-5210) for the most current information about the Ward Valley facility and any available options for handling LLRW beginning in 1993.

* * * * *

About the Authors:

Alan Pasternak, Ph.D., has served as the Executive Director of Cal Rad Forum since 1983, and coordinates Cal Rad's legislative, regulatory and public information programs. Dr. Pasternak served as a member of the California Energy Commission from 1975 to 1979, and has worked extensively on energy and waste management issues.

Jennifer Hernandez is a partner and co-chair of the Environmental and Land Use Practice Group of Graham & James. Ms. Hernandez serves as counsel to the Forum and to universities, hospitals and biotechnology companies which generate LLRW. Prior to returning to her private practice at Graham & James, Ms. Hernandez served as the senior environmental lawyer for the nine-campus University of California system. Ms. Hernandez' practice includes environmental compliance counselling on air quality; hazardous, medical and radioactive materials; project permitting for industrial and utility developments; and land use compliance and litigation.

For Further Information Contact the Authors at:

CALIFORNIA RADIOACTIVE MATERIALS MANAGEMENT FORUM
P.O. Box 1638
Lafayette, CA 94549-1638
510/283-5210

or

GRAHAM & JAMES
One Maritime Plaza, Third Floor
San Francisco, CA 94111
415/954-0286

Questions and Concerns

READING 2

As you read this article consider the following:

1. This reading refers to the Ward Valley dump that will take care of most of California's LLRW.

2. The letter to the editor indicates the level of misconception that exists. Can you list these misconceptions?

Reading 2

June 3, 1992
In response to Bruce Campbell's letter in the June 1992 Southern Sierran re: Ward Valley
Dear Editor:

It is an insult to the League of Women Voters and the Sierra Club members to assume they were duped. First of all, Ward Valley is a proposed low level waste site (LLWS) which means that a) only certain types and quantities of radioactive materials can be shipped there, b) the nuclear industry would be but one of the users of the facilities and c) the most important group using this facilities would be research institutes, universities, biotech industries and hospitals. This last group uses radioactivity as a vital means of discovering the cause of and in the treatment of diseases, especially cancers. Additionally, one of the suppliers of radioactively labeled drugs and biochemicals for these groups is a large employer in southern California who also needs some means of safely disposing of their low level radioactive waste.

The Ward Valley plan is not some quickly thrown together plan; it has been in the planning stage since 1980. There are approximately 10,000 pages of careful study and data regarding site selection compiled by the Department of Health Services. To discard this data just because it does not fit someone else's agenda is irresponsible and shortsighted. Also, the League of Women Voters have been involved since 1984 when public discussion of this issue began with the Sierra club, the Native American Heritage commission, county Board supervisors and local citizens. This Citizen's committee (which incidentally includes <u>no</u> elected officials, agency representatives or other professionals") has been actively participating in the whole siting process.

Ward Valley may be sacred Native American land, but via the Native American Heritage Commission, this group has been involved in the discussion of the site. Additionally, Ward Valley stretches some 350 square miles; the LLWS would occupy only 70 acres at the far north end of the valley.

The proposed site is some 20 miles from the Colorado River, but anyone with a topography map of the area will notice one obvious thing: the Dead, Sacramento, Chemehuevi, Piute, Turtle and Old Woman Mountains surround Ward Valley; specifically the Dead and the Sacramento Mountains lie between the proposed site and the Colorado River. Hence, the viewpoint that a flashflood through the site would contaminate the River (and LA's water supply) seems remote at best.

The desert Tortoise inhabit the area, but this issue has been carefully reviewed by the California Department of Fish and Game, who have suggested (and with which U.S. Ecology agreed) that special fences be installed to protect the tortoises from the access roads. If only so much thought had been given to the myriad of roads running through all the pristine natural area.

Let's understand some facts about radioactivity: Uranium 238 is a naturally occurring radioactive element in the California rock and soil. It is also the fuel used in the nuclear reactor, but no amount of uranium is allowed at low level waste facilities; in other words, <u>none will go to Ward Valley</u>.. What does go there from nuclear industry is low level and fairly short-lived isotopes. The claim that most of the waste is "gloves and booties" is absolutely true. Any time a glove or article of protective clothing is even possibly contaminated, it is disposed of as radioactive waste. So, imagine then, as a surgeon changes gloves and gown often to guard against infection, so do radiation workers change protective clothing to protect themselves and others. Because of the incredibly exacting safety standards which apply to nuclear facilities, hospitals and universities, this does indeed create a lot of waste in gloves and booties.

With regards to the safety of "unlined trenches" there are two issues to consider: First, a lined trench would mean that sealed barrels could potentially sit in wet conditions which, over a period of time, would greatly accelerate their corrosion. Secondly, the packing of these barrels must meet very rigid safety standards set by the state California Department of Health Services and the U.S. Nuclear Regulatory Commission. Additionally, every waste generator is responsible for that waste "from cradle to grave" (often with million-dollar indemnifications) which insures that the party packing the waste takes this responsibility very seriously.

Currently there are only 3 sites in the United States which take low level radioactive waste: Nevada, Washington and South Carolina. Understandably, they want the responsibility to fall where it should: in the generator's own state. To discourage shipment from other states, these sites are beginning to increase their rates (currently at $570 per 55 gallon barrel at the Washington site). Many states, including California, promised several years ago that they would have their own facilities by January 1, 1993, but since it is June 1992 and the site is still in limbo, California is reneging on its promise and, understandably, the existing facilities want to hike their rates even more.

Universities and hospitals in this state are already under economic distress. Additional funds needed to dispose of their radioactive waste may simply not be available, which means that high caliber research will move to other states or countries where their work can continue and hospitals may have to stop certain types of life-saving treatments.

Radioactivity should not be feared because people do not know what it is. It is an invaluable research tool. Most of the Nobel Laureates in Science and Medicine in the past decades have used radioactivity. Bone marrow transplants, now routinely used to save leukemia patients, would not have been possible without radioactive material used in the research to design and perfect the technique. Diseased thyroids are routinely treated with radioactive Iodine. Radioactivity is not something invented by modern science. It has been around since the earth was molten mass; however its irreplaceable role in the discovery of many aspects of chemistry, biology and medicine has only recently been defined. Rather than the NIMBY hysteria surrounding radioactivity, I suggest people judge the dangers to our environment as a whole.

I consider myself an environmentalist. I carpool, recycle, buy judiciously, restrict use of energy and water, pick up the trash I find on beaches and in natural reserves and am vocal to my government representatives when I sense irresponsible acts of others will harm the environment. I do not think that nuclear power is the ultimate answer to our energy needs, but considering the pollution caused by burning coal or oil, or the incredibly toxic arsenic waste created in the production of relatively inefficient and geography-limited solar panels, I don't believe that our current technology has any quick alternatives. Solid, well-thought out science, reviewed by informed citizens is a prudent policy and one which I believe has been followed here. Let's deal responsibility with what waste we create. And remember, when you turn on that light bulb, where does *your* electricity come from?

Susan Fisher-Young
Assistant Radiation Safety Officer
California State University Fullerton
800 State College Blvd.
Fullerton, CA 92670
(714) 773-2687
Home phone: (714) 524-1045

Questions and Concerns

READING 3

As you read this article consider the following:

1. Where are the Farallon Islands?

2. When did the dumping take place and what was dumped?

3. Is the problem one for the future or is it a reality today?

Radioactive waste site is under investigation

One of the West Coast's richest marine life breeding grounds – the waters around the Farallon Islands off San Francisco – has been a nuclear waste dump since 1946. A recent federal report has raised fears of radioactive contamination of commercial and sport fish catches from the area.

By Brooks Townes

Beginning in 1946, at least 47,500 barrels of U.S. atomic lab and industrial wastes were dumped into the sea adjacent to the islands and within the Gulf of the Farallons National Marine Sanctuary. The 55-gal. drums, unloaded near the islands until 1970, are corroding and may have burst from water pressure, according to the report prepared by the National Oceanic and Atmospheric Administration (NOAA).

Though the waste stems from as far back as the Manhattan Project, the government did not acknowledge it was out there until some 20 years ago. The effects of the leaking drums on marine life at the site have been largely unstudied, the NOAA report says.

The agency is now calling for studies on radioactive and toxic chemical contamination in the area's Pacific herring, Dover sole, rockfish, sablefish and Dungeness crab stocks. NOAA and area congressional representatives are asking Congress for $750,000 to explore the problem.

"Radioactivity is only one problem," says Ed Ueber, Farallon Island sanctuary manager. "There's also unknown amounts of all sorts of hazardous chemical wastes out there. How much waste and the amount of leakage are unknown."

Front-page news stories, however, have apparently not dampened local fresh fish sales, according to a random survey of area retail markets. Ueber says that the little sampling of Farallon catches that has been done has turned up only traces of higher contamination, "and you can say that the Ueber family eats the fish all the time."

Still, no one knows what and how much is leaking from the concrete-lined barrels. "It is an absolute fraud to say that some level of radiation is safe, " says Dr. John W. Gofman. "There is no safe level. At any dose, there's a risk [of cancer]. As doses go up, so does risk.

"It is a propaganda fraud to say that there is such a thing as 'below levels of concern' or 'permissible' levels of radiation that are safe."

Gofman is professor emeritus of medical physics at the University of California, Berkeley, a veteran nuclear scientist and a former commercial salmon troller. He is greatly disturbed by the dumping practices but thinks the proposed studies are largely a waste of time.

"It's out there. We know that. We

Courtesy of Brooks Townes and National Fisherman

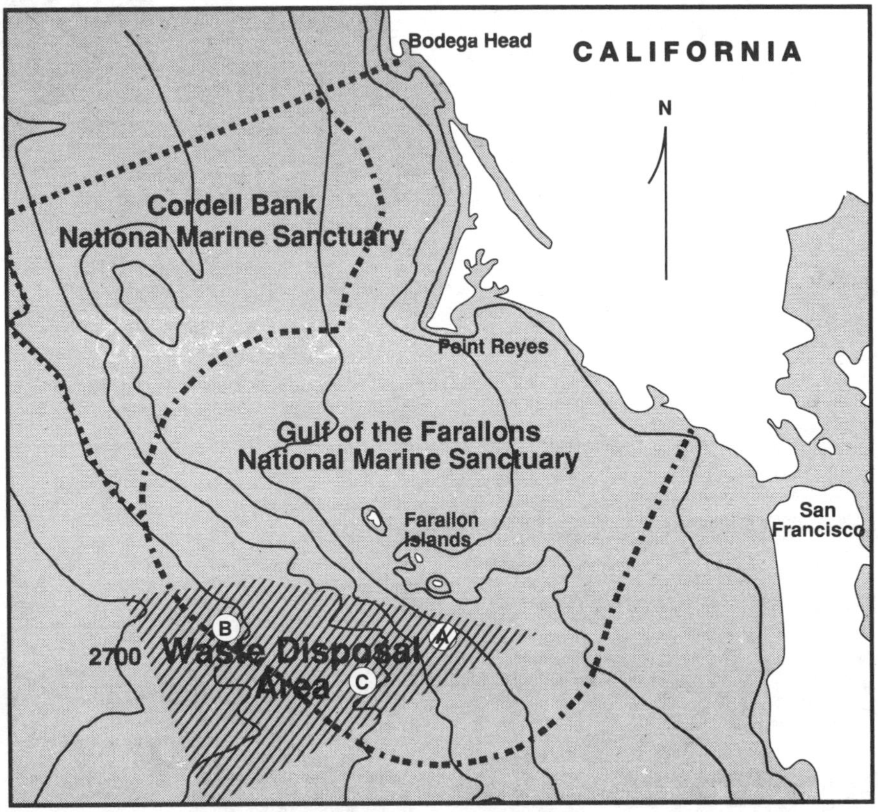

Materials containing radioactive wastes and heavy metals have been dumped at the disposal site – which is within the marine sanctuary area – for decades. So far, however, the Environmental Protection Agency has done nothing about it.

know what radiation does, and you can't clean up the Pacific Ocean," he says. "The dumping was done in an era of utter stupidity. It should be a lesson in how stupid we can be. Sure, eating the fish involves some risk, but there's no point in worrying about it now." However, he does think it's important to monitor the radioactivity and chemical pollution in fish caught from the area.

While it has been known for a couple of decades that radioactive material was dumped off the Farallons, little has been done to determine exactly what was dumped by whom or when. Two years ago, a contractor to the Environmental Protection Agency (EPA) urged the agency to rate the Farallon Island Radioactive Waste Dump as a "high priority" for investigation, using federal Superfund monies. But the EPA has taken no action.

Ignorance about the extent of the dumping stems from government secrecy about its atomic energy program. So far, NOAA researchers have learned that the drums were dumped by at least four private firms under contract to the Atomic Energy Commission. Besides the radioactive wastes, phenols, cyanides, mercury, beryllium and other heavy metals, plus explosives, garbage and dredge spoils have been dumped off the Farallons.

All these dangerous materials were enclosed in 16-gauge, second-hand steel drums, which are expected to last in sea water no more than 10 years. Many of the drums were lined in concrete, but even those were not expected to last more than 30 years.

According to the best-available information, the drums contained an estimated 14,500 curies of radioactive junk, which was reported to be 1,000 times the estimated radioactive release from the accident at the Three Mile Island nuclear plant in Pennsylvania.

Divers joke about locating urchin and abalone by the glow.

The barrels were dropped in depths ranging from 300' to 6,000' of water, and in a 1974 study, 25% of the barrels observed had already caved in from water pressure.

Plutonium 239 and 240, with a half-life of 24,360 years, and cesium 137, with a half-life of 30 years, were the most common contaminants found in the water and mud.

"No definitive studies have been conducted and thus it is difficult, if not impossible, to assess the degree of risk," says the NOAA report.

So far, the sediments at the Farallons have shown greater-than-normal levels — perhaps more than 1,000 times greater — of long-lived plutonium and cesium. Those dangerous elements can move up the food chain from worms and tiny shrimp that feed in the sediments and then are eaten by sole, rockfish and sablefish.

If it proves feasible to retrieve some of the barrels, scientists fear moving them may release more radiation. Resealing drums at those depths has never been successful.

So, while divers joke about locating urchin and abalone in the area by the glow, scientists face a tough and dangerous problem. □

Discussion for Hazardous Waste: Not in Our Backyard!

1. What sort of research would you suggest be done to understand the dangers of LLRAW?

2. If there are five levels or grades of radioactive waste, should an education program be designed to convince the general public that some levels of waste danger are acceptable?

3. Should the cost of the alternative, the 'do nothing scenario,' be publicized and made a part of the issue?

4. If you were the public relations chief of the Waste Disposal Site company what would you do? What sort of strategy could you design to solve the NIMBY problem?

5. Do we have 'rights' to prosecute those who polluted in the past? What if it was a legal dump site in the past?

6. The minimum permissible level of radiation is a controversial issue. What evidence is presented for both pro and con arguments?

Introduction

The Fight To Save Our Wetlands

This is a study of what several of the western states are doing with their wetlands. Are they doing a conscientious job in protecting them and how are they dealing with such problems as maintenance, protection of seasonal wetlands, governmental cooperation, restoration of degraded areas and mitigation measures?

Once scorned and considered worthless, wetlands of the United States are becoming more appreciated as their hydrological, physical, biological and chemical benefits are recognized. There is, however, an ever growing demand for agricultural land and areas for development and wetlands are viewed by some as the ideal place to start. The western states are looking for ways to satisfy both the environmentalists and developers and it is a tough assignment.

SECTION 404 OF THE FEDERAL WATER POLLUTION ACT

Protection of wetlands has fallen under the states jurisdiction ever since 1977 when Congress gave them the right to administer development permits in these areas. It was believed that state control would streamline the processing of development permits as well as protect the interests of the wetlands. Most states did not perceive this as being the case.

Wetland areas are generally rich habitats for wildlife, especially waterfowl and fish. In 1990 the Federal Energy Regulatory Commission (FERC) refused to issue an operating license to a hydropower and irrigation project along the Platte River in Nebraska after the agency was sued for their refusal to review environmental conditions. These problems were the result of up stream damming and water diversion which had taken place over the past hundred years. Some where along the North Platte (Wyoming) and the South Platte (Colorado) over 60% of the water has been siphoned out! This loss of water has jeopardized the habitats of the endangered Whooping Crane, Bald Eagle, and the Piping Plover. The FERC's decision to consider the effect of stream flow reduction on the habitats for endangered species is a positive step and the commission is becoming sensitive to environmental issues.

SEASONAL WETLANDS

A common misconception that is threatening the protection of many wetlands is the notion that the value of a wetland is directly correlated with how wet it is, how often it is flooded and for how long. One criteria that has been suggested for determining the value of a wetland is to see if it splashes when a duck lands in it. Several implications are to be drawn from this test and this type of wetland grading shows little understanding of overall wetland ecosystems.

Eighty percent of the San Francisco Estuary has been lost and 77% of what is left has been declared seasonal marshland by the Fish and Wildlife Service. This estuary provides shelter for 400 species of wildlife, 55% uses both the permanent and seasonal wetlands.

SOME STATES PERSPECTIVES

The governor of the gtate of Washington has developed a strategy to strengthen existing laws and regulations with his Executive Order 89-10. This order has a goal of no net loss of wetlands in acreage or function. Executive Order 90-04 gives more specific mandates to the state agencies in order to protect the state wetlands. It also created the Interagency Wetlands Reserve Board to monitor compliance with EO 89-10 and 90-04. A growth management measure passed in 1990 has established goals and policies for development in critical areas. These include wetlands, recharge areas, fish and wildlife habitats, frequently flooded areas and geologically hazardous area.

Alaska has the highest percentage of area designated as wetlands. In fact, 62% of the nation's wetlands are in this state. With the push to explore for oil and minerals this habitat has become threatened. In most cases where the tundra has been disturbed it is due to gravel being placed over it for road ways and equipment pads. When the gravel is removed the damaged area has been noted to recover, with a high reinvasion rate for the native species. Does this bode well for the saving of damaged Arctic wetlands?

Oregon was the first state to pass a planning law that protected its coastal wetlands. It remains stringent in its requirements for development of wetland areas, allowing only those that meet three criteria. First being that no practical alternative exists, the second, that the project will cause no significant degradation of municipal water supplies and wildlife habitat and the third being the requirement of steps to minimize any adverse impacts. Oregon has been very strict with development of tidal wetlands. Mitigation may mean restoration compensation at a ratio of 6:1, six acres restored for one developed. Most projects are checked for compliance and annual reports are required for up to five years following development. The state also enacted a Wetlands Mitigation Banking Act in 1987 to aid small (less than 5 acres) projects, but this bank can only be used when everything else fails.

Amy Fields

Reading 1

Questions and Concerns

READING 1

When reading this article consider the following:

1. Where do we find wetlands?

2. What are the guidelines for designating a wetland?

3. Can any one species be used as the indicator of a marsh?

Roundtable

Is wetter better?

. . . darlin' it's better, down where it's wetter, take it from me . . .
— Sebastian the Crab, *The Little Mermaid*, Walt Disney Productions, 1990

To those who are skeptical about our wetlands commitment, let me suggest you desist from merely caricaturing regulatory efforts and help us figure out how to ensure continuing habitat for ducks and other waterfowl.
—William K. Reilly, Administrator of EPA, addressing the American Farmland Trust, 7 March 1991

A current misconception that threatens to undermine existing wetland protection efforts is that the value of a wetland to society is positively correlated with increasing wetness, that is, increasing frequency, depth, and duration of flooding, inundation, or saturation. The hydrological variability among wetland types is the focus of those who question whether all wetlands should receive equal protection under federal law. Although ecologists agree that all wetlands are not physically, chemically, and biologically equivalent, there is no strong basis, in terms of relative ecological value, for comparing different types of wetland systems. Science fails to support the practice of assigning lower functional values to periodically saturated wetlands in comparison with permanently inundated or saturated wetlands.

by Robert A. Leidy, Peggy L. Fiedler, and Elisabeth R. Micheli

The public has only recently recognized wetlands as among the nation's most valuable natural resources. This recognition is based on research that demonstrates that, in addition to maintaining plant and animal biodiversity, wetlands perform other important functions such as food chain support, production export, water quality improvement, groundwater discharge/recharge, flood and erosion control, and global climate regulation. Ironically, it is the success of efforts to provide a sound technical basis for regulating wetlands that has allowed critics of federal wetland protection to seize upon an apparently ecological argument for reducing future levels of protection.

The belief that wetter-is-better underlies public perception that federal regulatory agencies have overstepped their congressional mandate under the Clean Water Act to protect wetland and aquatic habitats, and this belief is beginning to shape future federal wetland policies (Bohlen 1991). Critics argue that regulators have overreached their mandate by exerting jurisdiction over drier areas that either should not be defined as wetlands or at least should not be afforded the same degree of regulatory protection as the wetter wetlands.

For example, the "duck test" (or "splash test"), currently advocated by some politicians, landowners, and agency officials to assess whether a particular area is a wetland worth protecting, is paraphrased: If a duck lands in it and splashes, then the area must be a wetland (or, more precisely, a valuable wetland). Implications of the duck test are: ducks are the only appropriate indicators of wetland values, seasonally saturated wetlands are not important to ducks and other waterfowl, and seasonally saturated wetlands do not provide other significant habitat values.

The assumption that wetlands at

Table 1. Wetland habitats within the San Francisco Estuary, California. (Compiled from USFWS 1991.)

Wetland habitat type	Area (hectares)
Predominantly permanent inundation or soil saturation	
Mudflat*	25,949
Salt/brackfish/freshwater tidal marsh†	17,964
Active salt pond	14,852
Subtotal (% of wetlands)	58,765(23)
Predominantly seasonal inundation or soil saturation	
Diked marsh, vernal pool, abandoned salt pond††	34,467
Farmed wetland††	156,176
Riparian forest	5066
Subtotal (% of total wetlands)	195,709(77)
Total wetlands	254,474

* Estuarine intertidal, including rocky shore.
† Estuarine intertidal.
†† Palustrine.

the drier end of the moisture gradient have little value is inconsistent with current understanding of the ecology of wetland systems. Drier wetland types, such as riparian woodlands, vernal pools, and diked marshes, play a crucial role as links between upland and aquatic habitats and are, therefore, central to the overall health of both ecosystems. For example, the wetter-is-better notion ignores how wetlands with different hydrologic regimes create complex mosaics of critical wildlife habitat.

We attempt to dispel the myth of wetter-is-better by highlighting patterns of wildlife use within wetland habitats in the San Francisco Estuary. Our analysis, based on one measure of wetland functional value, illustrate that wetlands with seasonal hydrology are a least as valuable as those with permanent hydrology. We conclude that unfounded generalizations about the relatively low value of drier wetlands are simplistic, scientifically insupportable, and inevitably detract from serious efforts to maintain and restore the physical, chemical, and biological integrity of the nation's waters—a clear mandate of the Clean Water Act.

Characteristics of wetlands in the arid west

Wetlands[1] are typically situated within a landscape along moisture and topographic gradients between upland habitats and deepwater aquatic systems, although some wetlands may be hydrologically isolated from larger aquatic systems. Although all wetlands are characterized by the presence of shallow surface water, a high water table, and/or saturated soils, the timing, frequency, and duration of flooding or soil saturation varies considerably within and among wetlands.

Not all wetlands are easily recognizable, particularly in many portions of the arid West where annual rainfall is highly variable and distinctly seasonal. For some types of western wetlands (e.g., diked wetlands, vernal pools, farmed wetlands, and riparian forests), evidence of wetland hydrology may be obvious only during the wettest portion of the rainy season in average or above-average rainfall years and may only last for a few days to several weeks. Because of their irregular or intermittent hydrologic regime, these wetlands are popularly known as "seasonal wetlands." They should be more accurately termed seasonally flooded or saturated wetlands (i.e., they are wetlands year-round, but they exhibit flooding or soil saturation on a seasonal basis; Figure 1).

[1]The term *wetland*, as defined by the Environmental Protection Agency, Army Corps of Engineers, US Fish and Wildlife Service, and USDA Soil Conservation Service, is conceptually the same: each requires positive indicators of hydrology, hydrophytic vegetation, and hydric or saturated soils. The USFWS definition also includes as *wetland* nonvegetated areas such as mudflat, gravel beach, rocky shore, and sand bar (FICWD 1989).

The current controversy over how to identify wetlands focuses on seasonally flooded or saturated wetlands, which include wetlands where saturation to the soil surface is irregular, but the root zone of the prevalent vegetation remains within saturated soils (e.g., many riparian or forested wetlands). Observing the hydrology of seasonally flooded or saturated wetlands at a dry time of year might lead to the erroneous conclusion that the area in question is not a wetland, or, if a wetland, that it does not possess any significant functional values. Although the hydrologic regime in seasonally flooded or saturated wetlands is variable at several temporal and spatial scales, such variability should not lead to the conclusion that they exhibit no significant functional values. Given that different physical processes influence the hydrologic regime within each watershed, one would not expect similar functional values for permanently flooded or saturated wetlands and for seasonally flooded or saturated wetlands. Different, however, does not mean less valuable. In fact, the high value commonly attributed to permanently inundated or saturated wetlands may be contingent on their use by wildlife in conjunction with seasonally flooded or saturated wetlands.

Patterns of wildlife habitat use

Patterns of wildlife use in the San Francisco Estuary illustrate that individual wetlands with different hydrologic regimes combine to form a mosaic or complex of different wetland types, each with a critical role in support of the overall productivity of wildlife populations (e.g., for waterfowl, nesting habitat versus foraging or perching habitat). This example illustrates the significance of wetlands with seasonal hydrology to the health of the overall complex.

The San Francisco Estuary drains 40% of the land area of California and is the largest estuary on the west coasts of North America and South America. Eighty percent of the estuary's tidal wetlands have been lost or converted: a significant portion have been converted to diked and/or farmed wetlands with seasonal hydrology. Wetlands within the estuary total approximately 250,000 hectares (Table 1). Approximately 77% of all wetlands within the estuary have been classi- [illegible] d Wildlife [illegible] (USFWS [illegible] 3% are clas- [illegible] cterized by [illegible] nt inunda- [illegible] including [illegible] and active [illegible] th of these [illegible] are suscep- [illegible] ual varia- [illegible] based on [illegible] ht events.) [illegible] within the [illegible] t the drier [illegible] radient.

[illegible] species of [illegible] vertebrate [illegible] s, reptiles, [illegible] have been [illegible] y (USFWS [illegible] % of these [illegible] ded from [illegible] estuary's [illegible] ring some [illegible] Wetlands [illegible] drologic [illegible] ignificant [illegible] wildlife [illegible] bians and [illegible] nple, 50% [illegible] pecies are [illegible] oodland, [illegible] ally wet [illegible] less than [illegible] acreage [illegible] nds with [illegible] lso criti- [illegible] ng up to [illegible] t for spe- [illegible] otnote to [illegible] pulations

[illegible] Estuary [illegible] d habitat [illegible] descrip- [illegible] bitat rela- [illegible] he extent [illegible] than one wetland habitat type, reveals a complex pattern (Table 3). For example, as many as 48% of vertebrate wildlife species recorded using tidal salt, brackish, and freshwater marshes (i.e., wetlands with predominantly permanent wetland hydrology) also have been recorded using wetlands with seasonal wetland hydrology. Up to 55% of vertebrate wildlife species recorded from intertidal mudflat (i.e., permanently saturated wetland) also use some type of wetland with seasonal hydrology (not including riparian woodland). Thus, wetlands with seasonal hydrology are central to the wildlife habitat function of the estuary, a value generally ascribed to wetlands with permanent hydrology.

What about ducks?

Much of the recent debate has focused on use by ducks and other migratory waterfowl as the primary function of wetlands, which requires permanent hydrology. However, even using this narrow measure of wetland value, duck and migratory-waterfowl field inventories indicate the importance of wetlands with seasonal hydrology to these species. The case of the San Francisco Estuary again serves as a potent example.

Midwinter waterfowl censuses for the years 1984 to 1989 recorded between 100,000 to 340,000 individual birds from the San Francisco Bay and its adjacent wetlands (USFWS 1989). These figures constitute approximately 5–13% of all duck sightings in California's Pacific Flyway during that time period. Nineteen out of the 35 species of migratory waterfowl known to use the estuary were observed using the 140 seasonally saturated or inundated wetlands inventoried. Species used these habitats for foraging and roosting, especially during high tides and storms.[2] Thirty-five percent of waterfowl species observed in wetlands with seasonal hydrology were also recorded using two or more different seasonal wetland habitat types (USFWS 1991).

Thus, it is a misperception that wetland wildlife, including waterfowl, require wetlands with standing water for most stages of their life cycles. As demonstrated by observations in the San Francisco Estuary, wetland complexes (i.e., diverse groupings of wetlands exhibiting both permanent and seasonal hydrologic regimes in close proximity to one another) are critical to healthy waterfowl populations.

The value of wetland complexes and wetlands with seasonal hydrology as waterfowl habitat has been established elsewhere as well (Fredrickson and Reid 1986, 1987, 1990). For example, in managed impoundments in Missouri, optimal foraging conditions for the greatest diversity of waterfowl and shorebirds were found in wetland complexes characterized by predominantly shallow water depths (Fredrickson and Reid 1986). Most shorebirds preferred saturated habitats adjacent to ponded areas within these wetland mosaics. Similar seasonal shallow-water and/or irregularly saturated wetlands are found throughout the arid West and typically have a seasonal hydrologic regime. As, in the San Francisco Estuary example, these western wetland complexes support diverse waterfowl populations.

Recent developments in federal wetland policy

USFWS (1989) has identified habitats with seasonal wetland hydrology as a priority wetland habitat for acquisition for waterfowl within the San Francisco Estuary. Ironically, at the same time that seasonally saturated or inundated wetlands are considered to be of increasing importance, the federal government has taken actions that will result in significantly reduced regulatory jurisdiction over these areas. The recent removal from Clean Water Act jurisdiction of millions of hectares of wetlands that meet the

[2]Unpublished US Fish and Wildlife Service data, 1991, Sacramento Fish and Wildlife Enhancement Field Office, Sacramento, CA.

definition of "prior converted croplands" under the so-called Swampbuster provisions of the Food Security Act of 1985, as amended, effectively reduced federal protection for seasonally saturated wetlands.

The US Environmental Protection Agency, Army Corps of Engineers, Department of agriculture Soil Conservation Service, and USFWS recently proposed changes to the federal technical criteria for identifying and delineating wetlands. These changes (40 Code of Federal Regulations, Chapter 1, 14 August 1991), which would further reduce the scope of national wetlands protection, were proposed in response to criticisms from the Bush administration, Congress, and the public. These changes would essentially limit federal jurisdiction under Section 404 of the Clean Water Act to permanently inundated or flooded wetlands or wetlands that are saturated to the soil surface for prolonged periods, and they would effectively reduce or eliminate protection for many other valuable wetlands characterized by highly variable seasonal hydrology. The administration also proposes to establish a categorization scheme where only what they call high-value wetlands would receive existing levels of federal protection (OPS 1991). Wetlands at the drier end of the moisture gradient would receive lower levels of protection.

The corps' Nationwide Permit Program also assigns a relatively low value to wetlands with seasonal hydrology. The basis for issuance of one such permit (i.e., Nationwide

Table 2. Affinities of wildlife species for selected aquatic and wetland habitats within the San Francisco Estuary, California. (Compiled from USFWS 1991.)

Habitat type	Amphibians	Reptiles	Birds	Mammals	Total	Special status species*
Number of species (%)	14 (3.4)	30 (7.3)	286 (69.4)	82 (20.0)	412	110
Number (%) of species using listed wetland or aquatic habitat type						
Permanently flooded (aquatic)						
Open water	0 (0)	0 (0)	57 (19.9)	4 (4.9)	60 (14.6)	12 (10.9)
Lakes and ponds	6 (42.9)	5 (16.7)	76 (26.5)	9 (11.0)	94 (22.8)	12 (10.9)
Permanent inundation or soil saturation (wetland)						
Intertidal mudflat	0 (0)	0 (0)	48 (16.8)	3 (3.7)	49 (11.9)	4 (3.6)
Rocky shore	0 (0)	0 (0)	26 (9.1)	3 (3.7)	29 (7.0)	3 (2.7)
Tidal salt marsh	0 (0)	0 (0)	86 (30.1)	11 (13.4)	95 (23.1)	21 (19.1)
Tidal brackish marsh	2 (14.3)	3 (10.0)	173 (60.5)	17 (20.7)	192 (46.6)	22 (20.0)
Tidal freshwater marsh	5 (35.7)	6 (20.0)	143 (50.0)	22 (26.8)	174 (42.2)	22 (20.0)
Active salt ponds	0 (0)	0 (0)	85 (29.7)	0 (0)	82 (19.9)	11 (10.0)
Seasonal inundation or soil saturation (wetland)						
Diked seasonal marsh	4 (28.6)	3 (10.0)	55 (19.2)	13 (15.8)	72 (17.5)	10 (9.1)
Other freshwater habitats	9 (64.3)	10 (33.3)	56 (19.6)	24 (29.3)	98 (23.8)	18 (16.4)
Farmed wetlands	0 (0)	9 (30.0)	66 (23.1)	18 (22.0)	92 (22.3)	6 (5.5)
Riparian woodland	12 (85.7)	20 (66.7)	121 (42.3)	58 (70.7)	207 (50.2)	29 (26.4)

*Includes species designated by federal or state agencies as endangered, threatened, species of special concern, or extinct within the estuary. Includes invertebrates.

Table 3. Matrix of the occurrence of vertebrate wildlife species from various aquatic and wetland habitat types within the San Francisco Estuary, California. Numbers represent the proportion of species recorded from paired habitat types.* (Compiled from USFWS 1991.)

	Hydrologic regime										
	Predominantly permanent indundation or soil saturation						Predominantly seasonal inundation or soil saturation				
Habitat type	LP	IM	RS	TSM	TBM	TFM	SP	DM	FMW	FWH	RW
OW	0.62	0.27	0.20	0.48	0.63	0.57	0.67	0.17	0.18	0.13	0.02
LP		0.57	0.38	0.49	0.36	0.44	0.63	0.48	0.28	0.50	0.51
IM			0.39	0.87	0.73	0.67	0.80	0.39	0.55	0.43	0.06
RS				0.48	0.52	0.38	0.52	0.28	0.21	0.17	0.07
TSM					0.89	0.83	0.54	0.48	0.46	0.47	0.21
TBM						0.72	0.31	0.30	0.38	0.35	0.44
TFM							0.37	0.37	0.41	0.45	0.45
SP								0.35	0.43	0.35	0.10
DM									0.76	0.79	0.28
FMW										0.82	0.19
FWH											0.24

* Habitat abbreviations: OW = open water; LP = lakes and ponds; IM = intertidal mudflat; RS = rocky shore; TSM = tidal salt marsh; TBM = tidal brackish marsh; TFM = tidal freshwater marsh; SP = active salt pond; DM = diked marsh; FMW = farmed wetland; FWH = seasonal freshwater habitåts; RW = riparian woodland.

Permit 26) is that the filling of isolated waters, headwater streams, and their adjacent wetlands has no significant individual or cumulative impact on the aquatic ecosystem.[3] In the arid western United States, many high-value seasonal wetlands, such as riparian woodlands, occur as narrow bands of vegetation along intermittent streams. Fills of less than four hectares authorized by the Nationwide Permit 26 have the potential to result in the destruction of extensive lengths of streambed and adjacent riparian woodland. Large areas of other isolated, seasonal wetlands (e.g., California vernal pools that provide habitat for a variety of threatened and endangered plant and animal species, as well as migratory waterfowl and shorebirds), have similarly been destroyed under this permit.

The role of science in policy

Ecology provides the scientific underpinnings for most federal environmental legislation. Policymakers should be made aware that within the last decade a minor revolution, of critical importance to the conservation of nature and natural resources, has occurred in the theoretical arena of community structure and function. The classical paradigm—that landscapes remain at equilibrium if left unmanaged or undisturbed—has proven again and again to be problematic in its application. In response, the contemporary paradigm in ecology—patch dynamics—involves a number of shifts in research foci. One is the shift away from analyzing fine-scale phenomena and toward analyzing ecosystem- and landscape-scale patterns and processes. In addition, ecosystems are no longer assumed to display long-term stability in community functioning but rather are typified by episodic disturbances controlling community structure and function on these larger time scales. Thus, ecologists are now most concerned with how systems differ in behavior and at what scales and in what contexts their interactions are most intelligible.

New community theories that incorporate both natural and human-caused disturbances into the mechanics of the natural world should guide the management of natural resource systems such as wetland mosaics. The shift in ecological thought comes at a time when the most visible (and likely most crucial) environmental problems (e.g., tropical deforestation, rising sea level, global climate change, and loss of biodiversity) are best addressed by large-scale theory and planning. If the recent proposed changes in federal policy toward regulating wetlands are any indication, new advances in ecological theory are not being successfully paralleled in the implementation of policies protecting the nation's wetland resources. This divergence is unconscionable; policy makers must resist the path of least resistance to the resolution of natural resource conflicts, and scientists must insist that policy makers implement environmental legislation and policy that best conform to current understanding of natural systems.

Conclusions

Overall, it is apparent that a well-organized backlash against the federal wetlands-protection program is under way. Our concern is that when ecological arguments are used in these public policy debates they should reflect the actual state of understanding of wetland science. We must avoid environmental policies based on popular but poorly conceived notions about ecological systems. Although for Sebastian the crab, wetter may be better, rarely do such generalizations about habitat values have a place in national policy for protecting wetlands and other ecosystems.

[3] *Isolated waters* are generally defined as nontidal waters and their adjacent wetlands that are not part of a surface tributary system to interstate waters or navigable waters of the United States. *Headwater* streams are distinguished from those "above the headwaters," which the Code of Federal Regulations defines as nontidal waters with an average annual flow of less than 5 cfs.

Acknowledgments

We gratefully acknowledge Wayne R. Ferren Jr. and Thomas G. Yocom for their insightful comments and Paul Jones for contributions on the use of various wetland types by waterfowl.

References cited

Bohlen, C.C. 1991. Controversy over federal definition of wetlands. *BioScience* 41: 139.

Federal Interagency Committee for Wetland Delineation (FICWD). 1989. Federal manual for identifying and delineating jurisdictional wetlands. Cooperative technical publication of US Army Corps of Engineers, US Environmental Protection Agency, US Fish and Wildlife Service, and USDA Soil Conservation Service, Washington, DC.

Fredrickson, L.H., and F.A. Reid. 1986. Wetland and riparian habitats: a nongame management overview. Pages 59–96 in J.B. Hale, L.B. Best, and R.L. Clawson, eds. *Management of Nongame Wildlife in the Midwest: A Developing Art.* North Central Section Wildlife Society, Chelsea, MI.

_____. 1987. Waterfowl use of wetlands complexes. In *Managing Waterfowl Habitats: Breeding, Migration, Wintering.* US Fish and Wildlife Service, OIT, Fort Collins, CO.

_____. 1989. Wetland and riparian habitats: a nongame management overview. Pages 59–96 in J.B. Hale, L.B. Best, and R.L. Clawson, eds. *Management of Nongame Wildlife in the Midwest: A Developing Art.* North Central Section, Wildlife Society, Chelsea, MI.

_____. 1990. Impacts of hydrologic alteration of management of freshwater wetlands. Pages 71–90 in J.M. Sweeney, ed. *Management of Dynamic Ecosystems.* North Central Section, Wildlife Society, West Lafayette, IN.

Office of the Press Secretary (OPS). 1991. Fact Sheet: Protecting America's Wetlands. The White House, 9 August 1991.

US Fish and Wildlife Service (USFWS). 1989. Concept Plan for Waterfowl Habitat Protection: San Francisco Bay, California. North American Waterfowl Management Plan, category 27. USFWS, Region 1, Portland, OR.

_____. 1991. Status and trends report on wildlife of the San Francisco Estuary. Report Prepared for US EPA, USFWS, Sacramento Fish and Wildlife Enhancement Field Office, Sacramento, CA.

Robert A. Leidy is the science and field program manager and Elisabeth R. Micheli is an environmental scientist in the Wetlands Section, US Environmental Protection Agency, San Francisco, CA 94105. Peggy L. Fiedler is an assistant professor of botany and director of the graduate program in conservation biology at San Francisco State University, San Francisco, CA 94132.

Reading 2

Questions and Concerns

READING 2

When reading this article consider the following:

1. What power does an endangered or threatened species have?

2. Do the mitigation measures guarantee success of the threatened birds?

3. What are the errors of the past that make this such a tough problem today?

Federal Notes

FERC Attempts to Protect Platte River

John G. Sidle and Michael LeValley
Fish and Wildlife Biologists
U.S. Fish and Wildlife Service
Grand Island, Nebraska

John G. VanDerwalker
Executive Director
Platte River Whooping Crane Habitat Maintenance Trust
Grand Island, Nebraska

To protect the riparian habitat of several threatened and endangered species along Nebraska's Platte River, the Federal Energy Regulatory Commission (FERC) recently, and for the first time, imposed conservation restrictions on an annual license for the operation of a hydropower and irrigation project. In May of 1989, the court in *Platte River Whooping Crane Habitat Maintenance Trust v. FERC*[1] ruled that FERC's refusal to review environmental conditions for annual license issuance amounted to an abuse of its regulatory discretion. Following this ruling, FERC has not only accepted the advice of the U.S. Fish and Wildlife Service (FWS), the Nebraska Game and Parks Commission, and the Platte River Trust regarding the deleterious effects of two dam and diversion projects on downstream habitats, but also has amended one of the project's annual licenses to protect the endangered whooping crane *(Grus americana)* and its critical habitat, the endangered interior population of the least tern *(Sterna antillarum)*, the threatened piping plover *(Charadrius melodus,)* and the endangered bald eagle *(Haliaetus leucocephalus)*. Although FERC believed that it had no authority to amend the second project's license, it encouraged the project's operator to cooperate in providing appropriate instream flows for habitat and wildlife.

Under the Federal Power Act (FPA), FERC is required to issue an annual license upon the expiration of an existing license: annual licenses are granted each year until the project's new license is granted. Heretofore, annual licenses only authorized continued operations under existing conditions: modifications for wildlife conservation customarily were not allowed or considered. In February 1990, however, FERC reversed this tradition considering the effect that stream flow reductions have had on the threatened and endangered species that depend on the central Platte River and its riparian wetlands. In light of the Platte River projects' impending license renewal, as well as the coming renewal of countless other hydropower projects across the country, FERC's attention to habitat conservation during the issuance of this annual license bodes well for the inclusion of environmental protection measures in the future issuance of hydropower licenses.

A Dammed River

Like many other hydropower and irrigation projects in the United States, Platte River projects 1417 (Central Nebraska Public Power and Irrigation District) and 1835 (Nebraska Public Power District), were licensed originally with little, if any, consideration of fish and wildlife resources. Projects 1417 and 1835 were licensed for a 50-year and 44-year period, respectively, in 1937 and 1943 by FERC's predecessor, the Federal Power Commission. Project 1417 includes Kingsley Dam and Lake McConaughy reservoir on the North Platte River and a network of canals, offstream reservoirs and hydroelectric facilities for irrigation and electric power generation. Project 1835 includes diversion dams, offstream reservoirs, and canals for the diversion of water from the North Platte and South Platte Rivers for electric power generation and irrigation.

Downstream of these projects are essential habitats for the

1. 876 F. 2d 109, 19 ELR 20883 (D.C. Cir. 1989).

Order Rescinded

As the *National Wetlands Newsletter* was going to press, the Federal Energy Regulatory Commission (FERC) rescinded the interim conditions requiring a specified regime of mean monthly and daily flows that it recently imposed on two hydropower operations on the Platte River. Fortunately, the provisions requiring restoration of least tern and piping plover nesting habitat have not been suspended.

Although the FERC order of February 14, 1990, delivered a *mandate* to the Nebraska Public Power District (NPPD), it only sought the *cooperation* of the Central Nebraska Public Power and Irrigation District (CNPPID)—the controller of the vast majority of the available water storage—because FERC did not believe that it could directly amend the CNPPID's annual license. According to FERC, the CNPPID's license does not allow FERC to modify its operation by requiring it to use stored water for beneficial public purposes. And unfortunately, the CNPPID is apparently unwilling to consider seriously taking any voluntary actions that would help meet the original objectives regarding wildlife and wetland monthly flow levels.

Consequently, the NPPD has requested a stay of the order, asserting that without CNPPID's voluntary compliance with the order's flow regime, NPPD would deplete its own water supply. Therefore, given that FERC will not amend its order to *require* CNPPID to cooperate, the original objectives are not likely to be met for most of the year.

The United States Fish and Wildlife Service (USFWS) contends that the requirements of the Endangered Species Act are relevant in the issuance of these annual licenses. The USFWS believes that FERC should consult with the USFWS in order to ensure that its decisions on both annual and new licenses do not jeopardize the continued existence of listed species or destroy or adversely modify designated critical habitats. Importantly, the Federal Power Act and its regulations do not abrogate the explicit requirements of the Endangered Species Act (ESA) and its implementing regulations. FERC, however, believes that the issuance of an annual license is a nondiscretionary, ministerial act that does not, in itself, constitute an "agency action" under §7 of the ESA.

Aside from the annual licenses, there is still the need for new licenses for NPPD and CNPPID, and these licenses are likely to mandate instream flow and other conditions. However, it may be three to five years before FERC grants new long-term licenses incorporating the extensive environmental protection provisions that are needed *now* for the Platte River.

bald eagle, whooping crane, least tern, and piping plover. A 53-mile segment of the Platte River valley has been designated as critical habitat for the whooping crane. The Platte River valley provides whooping cranes with both feeding areas and nocturnal roosting sites, thereby improving the birds' physical condition for migration and later reproduction in the Canadian Northwest Territories. Wide river channels provide secure roosting sites for the cranes, and adjacent wetlands provide invertebrates, an important food source for the cranes. In addition, least terns and piping plovers nest on barren-to-sparsely vegetated sandbars in the River. The preservation of sandbars in the river channel and appropriate instream flows have been identified in the FWS recovery plans as essential habitat conservation measures for the recovery of these species.

However, up to 70 percent of the flow in the central Platte River has been lost since the mid-nineteenth century due to upstream damming and the diversion of water along the North Platte, South Platte, and Platte Rivers. Stream flow reductions have contributed greatly to current habitat conditions along the central Platte; a once wide and open river channel has been reduced to a narrow channel, flanked by a dense riparian forest that has replaced the riparian wetland habitat. Project operations resulting in diminished flows have reduced the availability of braided river channel and sandbar habitat for whooping cranes, least terns, and piping plovers, and forage fish for least terns and bald eagles.

Permit Renewal Process

The Electric Consumers Protection Act of 1986 amended the FPA to require that equal consideration be given to fish, wildlife, and other aspects of environmental quality in FERC's licensing proceedings. Indeed, §4(e) of the FPA now states that in deciding whether to issue a license.

> [T]he Commission, in addition to the power and development purposes for which licenses are issued, shall give equal consideration to the purposes of energy conservation, the protection, mitigation of damage to, and enhancement of, fish and wildlife (including related spawning grounds and habitat), the protection of recreational opportunities, and the preservation of other aspects of environmental quality.

Although FERC found the project's new license applications deficient, it did not set a timetable to correct the deficiencies and to undertake new license proceedings. Therefore, the FWS requested that FERC conduct formal consultation under §7 of the Endangered Species Act (ESA) to determine whether new conditions should be placed on the annual licenses of the Platte River projects.[2] The FWS indicated to FERC that the projects did, and would continue, to affect the bald eagle, whooping crane, least tern, and piping plover, as well as the whooping crane's designated critical habitat. Moreover, the FWS advised FERC that, under the ESA, it could not issue annual licenses without §7 consultation. FERC never responded to this FWS request. Thus in 1987, the Platte River Trust petitioned FERC to impose on the annual licenses conditions designed to protect the Platte River habitat of the whooping crane and other species.

FERC initially denied the Trust's petition because it believed the existing licenses contained no reservations of authority to amend an annual license for the protection of fish and wildlife or

2. The ESA requires that "each federal agency shall, in consultation with and with the assistance of the Secretary [of the Interior], insure that any action authorized, funded, or carried out by such agency . . . is not likely to jeopardize the continued existence of any endangered species or threatened species or result in the destruction or adverse modification of habitat of such species which is determined by the Secretary . . . to be critical."

other reasons. Although the Trust pointed out at a rehearing of the petition that project 1835's license did include an express reservation of FERC's authority to alter the license, FERC nonetheless denied the Trust's request for an assessment of the need for environmental protective conditions. FERC argued that it did not have sufficient information to determine appropriate mitigative conditions, and reasoned that during the *new* license proceedings, substantive information would be developed and Platte River conservation matters would be addressed.

The court of appeals held, however, that FERC had been informed of the substantive concern and information relating to the Platte River by FWS and private conservation groups since at least 1975. Therefore, the court stated ". . . it was irresponsible for FERC to ignore this [FWS] expert opinion and refuse even to conduct a preliminary investigation into this threat and the availability of interim measures to combat it."[3] The court went on to state that ". . . Congress expected FERC to exercise whatever authority it might have to introduce into *existing* licenses environmental protective conditions that in its judgement appear necessary.[4]

The court finally held that it was not reasonable for FERC to refuse even to explore or to consider the need for some interim environmental protection pending the completion of the relicensing proceeding.

FERC Imposed Conditions

Following this court ruling, FERC amended the annual license for project 1835 after concluding that "certain aspects of the Platte River habitat system have degraded and will continue to degrade pending relicensing as a result of the operation of the Districts' projects,"[5] and that river channel narrowing and vegetative encroachment have occurred as a result of project operation and will continue to occur absent interim mitigative measures. FERC required the project to maintain year-round instream flows in the Platte River based on the previous end-of-month volume stored in Lake McConaughy, and asked that project 1417 cooperate as well. At greater than 1.3 million acre feet (maf) in the lake, the instream flows, as measured at the Grand Island gauge, to be maintained are as follows:

- 1,100 cubic feet per second (cfs) mean monthly flow for the maintenance of bald eagle wintering habitat, forage fish habitat, and maintenance of wetland habitat for whooping cranes, from November 16 to March 22 each year;
- 2,000 cfs mean monthly flow during the whooping crane's spring (March 23 through May 10) and fall (September 16 through November 15) migration each year, for the maintenance of riverine whooping crane roosting habitat; and
- a mean monthly flow of 800 cfs and not more than 2,500 cfs during the interior least tern and piping plover nesting season (May 11 through September 15 each year) to provide dry nest sites, water barriers around nesting colonies to reduce predation and human disturbance, and to maintain forage fish.

Furthermore, when the previous end-of-month volume stored in Lake McConaughy is less than or equal to 1.3 maf and greater than or equal to 0.9 maf, a mean daily flow of approximately 400 cfs shall be maintained, regardless of the time of year.

FERC also required that eight permanent five- to ten-acre sites for interior least tern and piping plover nesting habitat be developed and maintained, safe from inundation during the nesting season, in the central Platte River. Implementation schedules and monitoring plans to evaluate the effectiveness of the instream flow regimes and nesting sites also must be developed by the project operators in consultation with Nebraska Game and Parks Commission, the Trust, and the FWS.

Settling the annual license issue for these projects signifies the beginning of a reconciliation of the needs of fish and wildlife conservation with those of power and irrigation. It is anticipated that additional habitat mitigation and conservation measures will be included in the final licenses for these projects. Moreover, following the court's ruling in *Platte River*, the FWS has directed its offices across the country to provide FERC with comment on any annual license where protective conditions are warranted. These advances on the part of FERC, the FWS, and the Trust are critical as increasing numbers of the nation's hydropower facilities, most of which were built in the 1940s, come up for reconsideration and license renewal in the 1990s.

---

3. 876 F. 2d at 117, 19 ELR at 20887 (emphasis added).

4. *Id.* 118, 19 ELR at 20888.

5. 50 Fed. Energy Reg. Comm'n Rep. (CCH) ERC 61, 180. The "Districts" are the dam operators.

Questions and Concerns

READING 3

When reading this article consider the following:

1. Can you have a wetlands in a permafrost zone?
2. Should the same rules apply to Alaska as to the rest of the states?
3. Note the methods for reintroduction of vegetation.

Reading 3

SCIENCE

Restoring Alaska's Wetlands

by Roger A. Post

Sixty-two percent of the nation's remaining wetlands acreage is in Alaska. The state's climate and remoteness have long protected its wetlands base, with losses only a fraction of those of the lower 48 states, but increasing oil and gas exploration and production have affected or threaten to affect more and more of the state's Arctic-tundra wetlands. The president's proposed National Energy Strategy and several bills introduced in Congress this year seeking to open up further expanses of Alaska to oil and gas exploration make paramount the question of the technical feasibility of restoring degraded wetlands on permafrost (rock or soil that continuously remain colder than 0 degrees Centigrade for two or more years). To date, there has been relatively little technical discussion or serious consideration of tundra wetlands restoration in policy arenas. The 1989 recommendations of the National Wetlands Policy Forum stated that "wetlands underlain by permafrost pose unique scientific challenges in deciding how to conduct successful wetlands restoration and creation. . . ."[1] This cautionary statement devolved to a pessimistic footnote in the 1990 memorandum of agreement (MOA) between the Army Corps of Engineers and the Environmental Protection Agency (EPA) concerning mitigation of wetland impacts. The footnote stated that "there are certain areas where, due to hydrological conditions, the technology for restoration or creation of wetlands may not be available at present, or may otherwise be impracticable."[2] Writing about the MOA, personnel from EPA and the Corps identified Alaska as an area where "permafrost conditions, hydrology, and climatic factors create technical problems that may make opportunities for wetlands creation and restoration not always practicable."[3]

Is pessimism about restoring or rehabilitating permafrost wetlands justified? I think not. The scientific literature supports a more optimistic view. Factors influencing long-term success of wetland creation and restoration in temperate regions have been identified in recent publications. In the absence of contrary evidence, it is reasonable to assume that these factors also apply to wetland restoration and rehabilitation in Arctic and sub-Arctic regions. Research on the effects of surface disturbance and on revegetation and rehabilitation of gravel fills and overburden piles in permafrost wetlands is providing new information on the potential for wetland restoration in the Arctic. Taken as a whole, this literature indicates that restoring or rehabilitating lands covered with dry, unproductive gravel fills to productive wetlands is technically feasible in permafrost regions.

Factors Affecting Restoration and Rehabilitation

In temperate regions, establishing and maintaining wetland hydrology are the most important factors determining initial success and long-term functioning of restored wetlands. A wetland's hydrology influences its vegetation composition and species richness. Naturally fluctuating hydrologic regimes may enhance primary productivity and maintain wetland communities that would otherwise change through ecological succession. Other beneficial factors include the presence of previously saturated (hydric) soils and plant propagules for establishing native wetland vegetation.

The quality of the landscape, in part, determines the presence of these beneficial factors. Landscapes little modified by humans, such as those in much of the Arctic, retain natural flow regimes and abundant sources of plant material for colonization of restoration or rehabilitation sites. In contrast, highly altered landscapes that have lost wetland storage capacity subject remaining wetlands to abnormally large runoff at some times and to diminished flows at other times. Propagules for native wetland vegetation may also be missing from such landscapes. For these reasons, wetland restoration in highly modified surroundings may require extensive hydrologic controls and planting or seeding programs to recreate previously existing wetlands.

Some wetlands are more easily restored than others. In a 1989 study for EPA, Jon Kusler and Mary Kentula ranked ease and probability of successful restoration for temperate wetlands in decreasing order as follows: estuarine marshes; coastal marshes; freshwater marshes along lakes, rivers, and streams; isolated marshes supplied by surface water; forested wetlands along lakes, rivers, and streams; and isolated wetlands supplied by groundwater.[4] These authors based their ranking on the reestablishment of natural hydrology, availability of plant propagules, growth rate of key species, control of water levels, and other factors. For example, reestablishing proper hydrology for the first three types of wetlands listed is relatively straightforward because they are within, or contiguous to, waterbodies that can provide natural hydrologic regimes. Reestablishing proper hydrologic conditions for the complex vegetation of isolated marshes is more difficult. Functioning forested wetlands with diverse vegetative communities are difficult to reestablish because they have exacting hydrologic conditions and trees grow slowly. Finally, reestablishing the hydrologic relationships for isolated wetlands supplied by groundwater may be even more difficult, and reinvasion of wetland vegetation could be slow because the waterborne transport of plant propagules is absent.

Relationships between wetland type and ease of restoration documented for temperate regions should apply to Arctic and sub-Arctic regions as well. The role of contiguous or adjacent waterbodies in establishing the hydrologic regimes of wetlands is based on physical principles and phenomena (e.g., tides, storm surges, seasonal flooding) that also occur in the North. Likewise, the availability of plant propagules for restoration of northern wetlands is determined by wetland distribution and the mechanisms present to transport propagules: wind, flowing water, and animals. Not only are these agents of transport present in Arctic and sub-Arctic regions, but northern peatlands are widely distributed, potentially providing many sources of plant material for the reinvasion of restored wetlands by native species. Finally, the distinct

Roger A. Post is a habitat biologist in the Alaska Department of Fish and Game, Fairbanks, Alaska.

Reprinted with permission of Roger A. Post, article originally appeared in National Wetlands Newsletter, Volume 12, Number 4

differences in growth rates of trees and marsh vegetation are as true in the North as they are in temperate regions and pose the same difficulties for establishing functioning forested wetlands.

Where do the permafrost wetlands of Alaska fit within this series? Vegetated estuarine and coastal wetlands have limited distributions in the continuous permafrost zone of northern Alaska and are largely unaffected by development at this time. In contrast, development of North America's two largest oilfields, Prudhoe Bay and Kuparuk River, has affected freshwater wetlands. These abundant, nonforested peatlands, supplied by precipitation and surface water and often influenced by adjacent waterbodies or seasonal runoff, are most similar to Kusler and Kentula's "freshwater marshes along lakes, rivers, and streams."[5] If the same general factors control the success of wetland restoration and rehabilitation in the Arctic as in temperate regions, rehabilitating Arctic-tundra wetlands should be only moderately difficult.

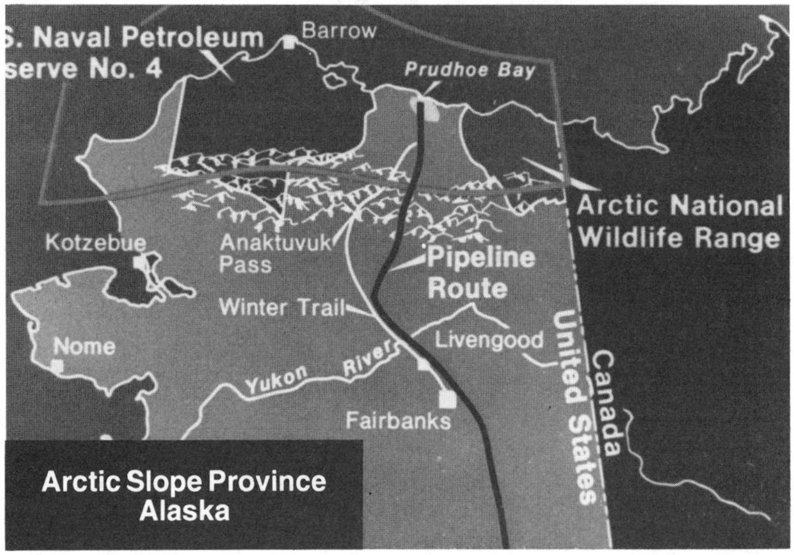

ARCO Alaska, Inc.

Potential for Restoring Tundra Wetlands

Several common characteristics of degraded tundra wetlands influence their potential for restoration. Wetland losses in Arctic Alaska primarily occur from placement of fill rather than from drainage. Oilfield development and transportation systems in areas of continuous permafrost require gravel fills from 1.5 to 2 meters thick to insulate soils that become unstable when they thaw. The Arctic Coastal Plain is a continuous mosaic of wetlands covering about 70,900 square kilometers. On it, gravel fills resemble islands and peninsulas in a sea of wetlands. Because large-scale hydrologic changes—such as might occur from agricultural drainage projects—are generally absent, the primary difficulty in wetland restoration—establishing and maintaining wetland hydrology—is not a concern.

In addition, gravel fills are usually placed directly on undisturbed tundra, taking advantage of the insulative qualities of the vegetative mat to reduce the fill thickness necessary to keep the underlying soils frozen. Hydric soils necessary for wetland restoration or rehabilitation remain intact. This characteristic of many degraded tundra wetlands means that *in situ* hydric soils can be exposed simply by removing the overlying gravel.

Compressed tundra exposed by removal of fill would thaw to a greater depth than undisturbed tundra, causing ground surfaces to subside as ground ice melts. Although fills kill and compress the vegetative mat, compression causes less thaw than does stripping vegetation away. Such thaw subsidence, or thermokarst, is common on the Arctic Coastal Plain; it is a natural process that often ponds water and initiates a cycle of physical change until thermal equilibrium is reestablished. This characteristic of degraded tundra wetlands not only implies that sites of fill removal would be wet but also that their physical and vegetative responses would parallel those of naturally occurring wetland types in the region.

Thick gravel fills in the Arctic are dry, sterile environments, which support little vegetation even after many years of disuse.[6] Such pads for oilfield facilities and reserve pits covered nearly 17 km^2 in the Prudhoe Bay Oilfield in 1983[7]—an area roughly the size of half of Manhattan Island. By 1987, pads in the Kuparuk River Oilfield covered an additional 8.4 km^2, and roads extended for 692 km—the distance from San Diego to San Francisco—covering more than 10.1 km^2 in the Prudhoe Bay and Kuparuk River oilfields.[8] Additional wetland fills on the Coastal Plain include overburden piles associated with gravel mines, a public highway, public airstrips, commercial pads occupied by the oilfield service industry at Deadhorse, and permanent settlements and their infrastructure.

Thick gravel fills on Alaska's North Slope can be modified and treated to support seeded grasses and to provide some benefit to wildlife, but such sites are not wetlands and provide few functions associated with the wetlands that they replace.[9] Where restoring or rehabilitating Arctic wetlands is the goal, the obvious first step is to remove abandoned or otherwise unused gravel fills. Removal is necessary because elevated gravel pads are hostile environments in the Arctic; they are wind scoured, do not trap or hold moisture, and may have naturally high salinity or be contaminated by salts or hydrocarbons associated with former drilling activities. In addition, plant species adapted to dry sites often do not occur in tundra wetlands and thus are unavailable for colonizing gravel fills. The chance of native vegetation colonizing unmodified, thick gravel pads, even at upland locations where dry-site colonizers potentially are available, is thought to be negligible.

The model of tundra disturbance by some researchers indicates that productive wetland communities will reestablish if changes to the original substrate are only moderate.[10] This would be the case following the careful excavation of fill to natural grade, even if a thin layer of residual gravel covered the compressed organic mat. Recent research on natural revegetation of abandoned fills on the North Slope has shown that fills less than 25 centimeters above the waterline established by the surrounding tundra are saturated and provide more favorable conditions for seedling establishment and growth than do thicker fills.[11] The most plausible assumption about the response of previously buried tundra is that moderate thaw subsidence would create depressions in the surfaces of rehabilitated sites, trapping runoff and

ARCO Alaska, Inc.

One of many North Slope facilities built to process and treat oil from the drill pads before piping it to the Trans-Alaska Pipeline System. Background shows permafrost wetlands.

precipitation and forming a mosaic of wetland types. These sites would then evolve toward eventual physical stability and thermal equilibrium.

Removal of residual gravel after careful excavation of fill to natural grade may or may not be advisable depending on desired characteristics of the rehabilitated or restored sites. Recent stream- and tundra-restoration work on the North Slope has demonstrated that specially modified vacuum trucks used for cleaning up oil spills can also remove gravel from tundra with very little damage to the surface, exposing the original vegetative mat and hydric soils. For sites with strongly compressed organic mats, removing residual gravel would reduce site elevation below natural grade, potentially ponding water and increasing thaw subsidence. Ponded water may support vegetative reproduction of emergent sedges and grasses but provides no surface for seedling establishment. In contrast, leaving residual gravel or removing residual gravel and reestablishing natural grade with fine-grained, organic-mineral soil (available from existing overburden stockpiles, particularly in the Kuparuk River Oilfield) may speed revegetation by exposing saturated soil surfaces for seedling establishment. These surfaces would then gradually subside from thaw consolidation of permafrost soils and acquire the wetland hydrology of the surrounding landscape.

Surfaces at natural grade and depressions formed by moderate thaw subsidence at sites of fill removal would exhibit the same hydrologic conditions as the undisturbed wetlands that surround them. In the Arctic, precipitation accumulates as snow for much of the year and rapidly melts at the beginning of the short summer. Snowmelt largely covers the tundra of the Arctic Coastal Plain with water, filling ubiquitous natural thaw ponds and lakes. Later, evapotranspiration generally lowers water levels in an annual cycle. This pulsed hydroperiod, characteristic of many wetland systems, including Arctic fill-removal sites, is widely believed to promote wetland restoration, maintenance, and productivity.

Likewise, wetland vegetation favorable to restoration or rehabilitation surrounds most wetland fills on the Arctic Coastal Plain. This vegetation can provide plant propagules for revegetation of sites where fill has been removed. Sedges, in particular, naturally revegetate saturated (thin) gravel or organic-mineral soils and can invade from adjacent tundra by vegetative reproduction over relatively short distances. Water and ice can also transport stem fragments of pendant grass (*Arctophila fulva*), which root from stem nodes to colonize new sites over greater distances. Although seed production has often been characterized as deficient in the Arctic, large amounts of seed are produced in years with favorable weather. Creating site conditions conducive to establishment and growth of native wetland plants is one of the central challenges in facilitating Arctic wetland restoration and rehabilitation.

Natural reinvasion of abandoned fills in Arctic wetlands is slow. Live plant cover at unrehabilitated sites (gravel or organic-mineral soils) with fill depths less than or equal to 25 cm increases at a rate of about 2 percent per year, implying that decades are required to reestablish good vegetative cover, even on saturated fills.[12] Nevertheless, applying basic restoration and rehabilitation techniques—such as reestablishing original site contours by removing fill, improving nutrient status, and supplying plant propagules—should enhance the development of functioning wetland communities. Recent research has shown that artificial fertilization speeds natural recovery of disturbed sites, which can also be successfully revegetated with cultivated varieties of native grasses used by geese and caribou. However, the grasses inhibit invasion by the diverse community of native forbs characteristic of the surrounding tundra, offering the short-term benefit of available grasses at the expense of a slower growing but broader mixture of native plant species that can support more diverse wildlife. Other experimental programs either proposed or in progress include transplanting aquatic grasses and sedges, harvesting and sowing native seeds, and applying organic-mineral overburden to control subsidence of disturbed sites.

Models of tundra responses to disturbance, based on empirical observations on Alaska's North Slope, can predict the success of wetland restoration or rehabilitation. Tundra response is dictated by original site characteristics (e.g., ground ice, soil, slope) and the degree of tundra disturbance associated with placement of fill, its subsequent removal, and the effect of restoration or rehabilitation treatments on site characteristics (e.g., thermal regime). Since disturbances ranging from no alteration to moderate alteration of tundra substrates produce either the original wetland community or a community at least as productive as the original, restoration or rehabilitation of marginally to moderately disturbed wetlands appears technically feasible using current technology.

Severely disrupted sites, however, such as those that would occur if the vegetative mat and soil were removed, have permanently altered thermal regimes that may prevent the sites from recovering to become communities as productive as those originally present. Severely disrupted sites thus pose greater challenges for restoration or rehabilitation than do moderately disturbed sites. Without remedial treatment beyond simple fill removal, severely disturbed sites may evolve by entering the natural thermokarst or thaw lake cycle, which has an approximate time scale of 1,000 to 10,000 years[13] and which carries the sites from dry or moist tundra to pond or lake formation and back to moist or dry tundra. The cycle begins when impounded water initiates thaw with gradual deepening of basins and changing of vegetation from aquatic sedge (*Carex aquatilis*) and cottongrass (*Eriophorum angustifolium*) to pendant grass. Thermokarst ponds may further deepen to open-water thaw lakes, which eventually intersect drainages and partially empty to create "basin-complex" wetlands, a habitat with high wildlife values. Most other stages of the thaw lake cycle also provide good habitat for waterbirds. Thus, even if severely disrputed sites are not treated sufficiently to meet a strict criterion for wetland rehabilitation (i.e., a community at least as productive as that originally present), they evolve in a way that emulates the natural environment and provides fish and wildlife wetland values.

Recommendations

Policymakers should carefully examine the scientific basis for statements concerning the technical difficulty of restoring or rehabilitating Arctic-tundra wetlands. The Arctic Coastal Plain is dominated by a pulsed wetland hydrology, supports a mosaic of wetland vegetation, and is underlain by hydric soils. These wetlands are on permafrost—which inhibits drainage—and have low rates of evapotranspiration, which maintain saturated soil conditions. Fills in Arctic-tundra wetlands may be easier to return to a wetland state than those in other environments because disturbance (e.g., fill placement and removal) usually causes thaw subsidence of the ground surface, increasing site wetness. Restoring permafrost wetlands in the Arctic thus does not appear demonstrably more difficult than restoring wetlands elsewhere and, in fact, may be easier.

Challenges, of course, remain. Wetland restoration or rehabilitation designed to recreate specific functions is more difficult than merely changing a dry gravel fill to a wetland. Wetland scientists need to fine-tune restoration technology to reproduce specific wetland communities and to influence physical evolution and ecological succession of rehabilitated sites. The need for research to achieve these goals should not be construed as a technical inability to convert sterile fills into productive wetlands. Rather, it should be understood as a continuing effort to minimize the impacts of economic development on the natural environment. Because resource extraction in the Arctic is a continuing and likely increasing activity, judicious removal and resue of gravel fills that are no longer needed will reduce the impact of such development on wetland communities. Recycling these fills will facilitate scientific studies of wetland restoration and rehabilitation on permafrost soils, return former wetlands to productive use by fish and wildlife, and reduce the need to open new gravel mines with their associated impacts on wetland habitats. ■

ENDNOTES

1. National Wetlands Policy Forum. 1989. *Recommendations For Federal Wetlands Legislation*. The Conservation Foundation, Washington, D.C.

2. *Memorandum of Agreement Between the Environmental Protection Agency and the Department of the Army Concerning the Determination of Mitigation Under the Clean Water Act Section 404(b)(l) Guidelines*, February 7, 1990.

3. Wilcher, L.S., and R.W. Page. 1990. "A Victory on Many Fronts." *National Wetlands Newsletter*. Vol. 12, No. 2.

4. Kusler, J.A., and M.E. Kentula, eds. 1989. *Wetland Creation and Restoration: The Status of the Science*. Environmental Protection Agency, EPA/600/3-89/038a, b. 2 Volumes.

5. Id.

6. McKendrick, J.D. 1986. "Final Cleanup at Selected (1975-1981) Wellsites, Sampling and Testing of Waters and Bottom Muds in the Reserve Pits and the Recording of Tundra Responses on the National Petroleum Reserve in Alaska (NPRA)." *Volume III: Recording of Tundra Plant Responses*. Prepared by Nuera Reclamation Co. (Anchorage) for the U.S. Geological Survey.

7. Walker, D.A., D. Cate, J. Brown, and C. Racine. 1987. *Disturbance and Recovery of Arctic Alaskan Tundra Terrain: A Review of Recent Investigations*. U.S. Army Cold Regiment Research and Engineering Laboratory Report 87-11.

8. U.S. Fish and Wildlife Service. 1987. *Comparison of Actual and Predicted Impacts of the Trans-Alaska Pipeline System and Prudhoe Bay Oilfields on the North Slope of Alaska*. Draft Report. U.S. Fish and Wildlife Service, Fairbanks Fish and Wildlife Enhancement Office.

9. Post, R.A. 1990. *Effects of Petroleum Operations in Alaskan Wetlands: A Critique*. Alaska Department of Fish and Game, Habitat Division. Technical Report 90-3; Jorgenson, M. T., T. T. Heffernan, and B.K. Lance. 1990. *Land Rehabilitation Studies in the Kuparuk Oilfield, Alaska, 1989*. Prepared for ARCO Alaska, Inc. and Kuparuk River Unit by Alaska Biological Research, Inc., Fairbanks, Alaska.

10. *Supra* note 7.

11. Jorgenson, M.T., et al., *Supra* note 9.

12. Id.

13. Carter, L.D., J.A. Heginbottom, and M. Woo. 1987. "Arctic Lowlands." *Geomorphic Systems of North America*, W.L. Graf, ed. Geological Society of America, Centennial Special Volume 2. Boulder, Colorado.

Questions and Concerns

READING 4

This article shows how citizen groups are urged to undertake projects to protect their environment. Can you see where Oregon's laws have developed from such an undertaking.

Reading 4

EPA/600/R-92/150
August 1992

An Approach to Improving Decision Making in Wetland Restoration and Creation

Authors:
Mary E. Kentula[1]
Robert P. Brooks[2]
Stephanie E. Gwin[3]
Cindy C. Holland[3]
Arthur D. Sherman[3]
Jean C. Sifneos[3]

Editor:
Ann J. Hairston[3]

[1]U.S. Environmental Protection Agency
Environmental Research Laboratory
200 SW 35th Street
Corvallis, OR 97333

[2]The Pennsylvania State University
Forest Resources Laboratory
University Park, PA 16802

[3]ManTech Environmental Technology, Inc.
USEPA Environmental Research Laboratory
200 SW 35th Street
Corvallis, OR 97333

Project Officer • Mary E. Kentula • Wetlands Research Program
U.S. Environmental Protection Agency • Environmental Research Laboratory
200 S.W. 35th Street • Corvallis, OR 97333

Printed on Recycled Paper

Volunteers and Natural Resource Monitoring

By Neal Maine

Editor's Note: We are pleased to offer the following special section as an example of how a monitoring program can be implemented by an agency despite limited staff and funding. Author, Neal Maine, exemplifies the positive impact one person can have on protecting coastal resources. He won the prestigious Chevron Conservation Award in 1988 after being chosen to represent Oregon by then Governor Neil Goldschmidt because "he educates Oregonians about our coastal resources by encouraging citizen involvement in conservation projects." Neal's contribution spans 26 years as a teacher in the Seaside Schools and as an instigator and volunteer in north coast wetland conservation projects. Asked how he has managed to achieve so much, Neal says: "When I see a need or where something is being threatened and nobody is doing anything about it, I just try."

Volunteers and Environmental Protection

Protection of the environment directly affects quality of life values. For this reason, environmental issues are often the focus of community action. Whether it is a group effort to preserve habitat for a valued species or to clean up a polluted area, many people are becoming more involved in protecting and restoring their environment. Federal and state agencies and nonprofit organizations promote involvement by volunteers in environmental activities because they know how essential local support is to the ongoing success of any project. Local participation includes activities such as community education, helping with work projects, building trails, recording observations, and "adopting" resources as varied as trees, streams and animals.

- In the state of Washington, the Adopt-A-Stream foundation invites volunteers to adopt a stream or wetland, collect data about the resource, and report the data to a planning organization for use in making decisions about needs for resource management.

- Volunteers work with the Oregon Department of Fish and Wildlife to conduct salmon spawn surveys, rehabilitate streams, and incubate salmon in streamside hatchboxes to help sustain Northwest salmon as both food and game fish.

- The National Wildlife Federation encourages individuals and groups to assist with winter bald eagle counts to ensure that the eagle population is not diminishing.

- The Nature Conservancy invites individuals and groups to participate in stewardship programs at selected reserves.

- Cannon Beach, Oregon, community volunteers educate visitors to the Haystack Rock tidal pools so they can enjoy the area without harming the pools or their inhabitants.

- The Environmental Protection Agency (EPA) has just completed a program called Streamwalk in Region 10 in which volunteers work with scientists to collect general information about a stream by mapping surveyed areas, characterizing the stream, and determining the cause of any adverse conditions found in the stream.

All of these programs involve citizens in the process of monitoring and protecting natural resources. When local citizens become involved with natural resources and their management, a new kind of protection emerges. The awareness of the entire community is focused on the problem at hand, ***the*** environment becomes ***our*** environment and is afforded the respect and protection reserved for the Earth as our shared home.

Monitoring Mitigation Projects

Funding for wetland mitigation, under Section 404 of the Clean Water Act, is based on the need to meet the legal requirements of proposals by developers, agencies, municipalities, and in some cases, individuals, which involve wetland alteration. Support for on-site monitoring of created or restored wetlands is often minimal. It is usually limited to single annual visits to the site for subjective evaluation. The collection of quantitative data is seldom required, although resource managers know that if they are to answer broad questions about the success of mitigation in restoring wetlands or specific questions about individual project success, the collection of valid data must be an integral part of all projects. A major limitation for including detailed on-site environmental monitoring in mitigation plans is the cost of professional services for collecting field data. Logistic problems such as transportation and scheduling constraints can also limit the collection of data on daily, weekly or monthly cycles. This is particularly true when sites to be monitored are some distance from qualified scientific staff. Providing opportunities for participation by volunteers can account for significant cost savings in professional services and travel expenses. More important, while helping collect specific data for a research project, volunteers begin to understand research goals and gain insight into the biological processes that occur in both created and natural wetlands. A local group of educated wetland supporters is created along with the wetland.

Aldo Leopold addressed this type of research 50 years ago in an essay in the Sand County Almanac on the development of the land ethic. He stated, "The last decade, for example, has disclosed a totally new form of sport, which does not destroy wildlife, which uses gadgets without being used by them, which outflanks the problem of posted land, and which greatly increases the human carrying capacity of a unit area. This sport knows no bag limit, no closed season. It needs teachers, but not wardens. It calls for a new woodcraft of the highest cultural value. The sport I refer to is wildlife (resource) research." Leopold concluded that, "The more difficult and laborious research problems must doubtless remain in professional hands, but there are plenty of problems suitable for all grades of amateurs."

The following sections discuss the concept of supporting an increased level of community involvement in scientific research activities, one in which local volunteers become an integral part of the effort. The discussion centers around how to build a research team that includes volunteers, using as an example the Trail's End wetland creation project in which six amateurs became part of a wetlands research team in Seaside, Oregon. Wetland research is the focus of this particular volunteer project, but local volunteers can be successful coworkers in most monitoring efforts.

The Trail's End Project

Comparing monitoring data from natural wetlands and wetland projects enables scientists to set standards for restored or created wetlands based on the quality that is attainable in that region and ecological setting. For example, scientists can identify when developing plant communities do not match the natural assemblages, and recommend corrections to project design based on typical local wetlands. The Trail's End study, conducted in the spring of 1989 in northwest Oregon, was designed to use volunteers to collect extensive data for a detailed monitoring study of a 15 acre created wetland, Trail's End, and three natural marshes. One of the objectives of the project was to test the use of volunteers for collecting scientific data within the context of a wetlands research project. Specifically, the objective directed researchers to evaluate the use of citizen volunteers in many aspects of the data collection.

Six volunteers from the Seaside community were recruited and trained to work with scientists on contract to the Environmental Protection Agency. These local volunteers accepted the responsibility for collecting data on vegetation, water chemistry, emergent and benthic invertebrates, hydrology, soils, and bird use. They also participated in mapping the sites, developing a photo record, and testing methods.

We learned from the Trail's End experience that the process of putting together the local team should receive special attention. This team is the sponsoring organization's link with the data collection activities and the wetland resource. The final accuracy of the data, and therefore, the reliability of the data, depend on the effectiveness of the initial team-building and training effort.

Team Building

Individuals who choose to become volunteers are the key to the success of a project. They are asked to put in many hours of work, often for long periods of time on a given day, and to collect data of a quality that will meet the standards of the sponsoring organization. In many cases, individuals are already involved with issues, activities, or management related to wetlands and wetland creation. These individuals are the best candidates for volunteers and for the role of Local Team Leader.

From the initial contact with the local community through the close of the project, every effort should be made to create a cooperative working environment in which volunteers and scientists are part of a real team effort. To do this, sharing of information from goals and objectives to the final evaluation of the project and possible use of the results is essential. Volunteers can be involved in planning their part of the process, or in making decisions about possible changes; they often have creative suggestions and superior knowledge of local resources.

We used a model with four interrelated roles in which volunteers and scientists work cooperatively. Each role demands certain responsibilities and desired attitudes and skills:

Principal Investigator

- Represents the sponsoring organization
- Responsible for the project's success

- Proven expertise in the design, execution, and management of major research projects
- Thorough knowledge of the specific wetlands under study
- Commitment to the concept of using volunteers in a research project
- Good communication and interpersonal skills

Off-Site Team Leader

- Represents the Principal Investigator
- Liaison between the scientists and the volunteers
- Responsible for on-site procedural decisions
- Thorough knowledge of wetland ecology
- Experience in identifying wetland flora and fauna
- Good communication and interpersonal skills

Local Team Leader

- Represents the volunteers
- Liaison between the Principal Investigator and the community
- (CRITICAL) Experience in organizing volunteers for research project
- Leadership skills for organizing and maintaining a productive and cooperative team
- Commitment to the success of the project and eagerness to learn
- Good communication and interpersonal skills for working with local agencies and the community
- Sound understanding of basic ecological concepts
- Knowledge of wetland flora and fauna

Volunteers

- Should be from the local area
- Willing to make a major commitment to the project
- Interest in protecting wetlands, plants and wildlife
- Some knowledge of natural history and science
- Eagerness to learn how to perform assigned tasks
- Ability to attend to detail and adhere to data collection schedules
- Ability to participate cooperatively in a team effort

The Principal Investigator and the Off-Site Team Leader are chosen by the sponsoring organization. The Local Team Leader and the Volunteers must be chosen with concern for both commitment and ability.

Local Team Leader

Following the designation of the Local Team Leader by the Principal Investigator and the Off-site Team Leader, a briefing meeting should be held to give the on-site person a full sense of the project scope and the general nature of the methods. It is critical at this time to establish the levels of responsibility for the sponsoring agency, the Local Team Leader and the volunteers. The Local Team Leader, who must be knowledgeable about community resources, will make personal contacts to recruit volunteers. This individual's clear understanding of his/her role in providing leadership by coordinating the data collection effort is essential for the success of the volunteer effort. It is also essential for the eventual validation of the field data collected and therefore, the reliability of the project results. An important factor in the success of the Trail's End project was the Local Team Leader's skillful leadership, planning and organization in determining the responsibilities of the volunteers and supporting them in their efforts.

Volunteers

The Local Team Leader contacts individuals who might be interested in volunteering for a community-based research project. It is helpful if they have had some personal association with research, although not necessarily as a professional scientist. These individuals are often found teaching science in schools, studying science in college, and volunteering in natural history programs. Others have had previous experience in a science-related profession, but are now working in different careers, and some have never had professional or academic training in science, yet have developed their skills through personal interest. The most critical attributes of successful volunteers are commitment to the success of the project; ability to attend to detail, adhere to data collection schedules, and work cooperatively; and eagerness to learn.

The Local Team Leader develops a list of the prospective volunteers. After the scope of study has been clearly defined, the Off-site Team Leader works with the Local Team Leader to match the backgrounds and skills of the volunteers to the specific study tasks. Personal contact is generally the most effective way to make the final decisions. An open meeting in the community stimulates interest and gives both team leaders a chance to meet with all of the potential volunteers. The meeting should be structured so that there is time for volunteers to introduce themselves, share their reasons for wanting to participate in the project, and work together in planning the next meeting around their needs for formal instruction.

Both initial and closing interviews are an integral part of evaluating use of local volunteers in data collection. Interviews with the research volunteers who participated in the Trail's End project revealed that they were interested in participating because of the chance to be a part of doing science, not studying about it; a personal interest in wetlands and natural history; the opportunity to use their skills and knowledge in a wetland setting; personal contact with professional researchers; and the excitement of being part of a study on local natural resources. By the close of the project, these volunteers were also concerned that without ongoing monitoring, local wetland mitigation projects would not meet the public need for protecting wetland values. Three of the volunteers at Trail's End were science teachers who transferred their skills to their high school students. The students are continuing to collect long-term data using the field techniques from the original study.

Training for Volunteers

After the team is identified, the next step is to complete the training that prepares the volunteers and the Local Team Leader to collect date at a level that meets the standards for the study established by the Principal Investigator. The data not only have to be accurate, they also must meet the Quality Assurance (QA) standards of the sponsoring agency (EPA in the case of Trail's End). The introductory training session is the first occasion on which all participants meet and is a good opportunity to begin building the team concept into the project. Encouraging everyone to contribute ideas and ask questions, and making decisions by consensus when appropriate, are two techniques for letting the volunteers know they are important to the project's success. In the first phase of the Trail's End project, presentations by scientists on the research objectives of the project, wetland ecology, data collection methods, correct use of reporting forms, and a discussion of the components of QA helped volunteers understand both the broad view of the project and the specific needs of the sponsoring agency. Training sessions were informal, with lots of time for volunteers to ask questions about the project and to clarify any personal concerns about participation.

Following initial training sessions, participants should be taken to the field where they can get hands-on experience with each of the procedures to be used in the project. For the Trail's End project, EPA staff presented instruction about a task, then volunteers performed the task while scientists played the role of QA observer. For example, using protocol instructions developed for this study, volunteers conducted water quality tests for oxygen and pH. The objectives were to compare water quality at the created wetland with the three natural wetlands and to test the applicability of the chemical field kits. Results were used to check the comparability of data between individuals and replicates. Each volunteer had an opportunity to work through the procedure and record data on the appropriate form. They received immediate feedback on the accuracy of the data. The training staff was sensitive to the concerns of the volunteers, who were allowed to repeat a technique until they were confident in its use. All volunteers agreed that this was the most critical part of the training. The hands-on experience enabled them to master the techniques and understand the expectations of the EPA staff. Volunteers also had the opportunity to change roles, from data collector to QA observer.

In some cases, such as the water quality sampling, data collection depended entirely on the skills and knowledge of the volunteers. In others, the volunteers supported the scientific staff by recording data or helping with equipment. After completing training, volunteers selected areas for which they would be accountable for data collection during the study. They then worked with the Off-site Team Leader and the Local Team Leader to establish a monthly schedule with a built-in QA cycle. As volunteers collected their data, the completed data sheets were given to the Local Team Leader to copy and file, with the originals going to the Off-site Team Leader for permanent project records.

Support for Volunteers

Ensuring proper training and a positive atmosphere for the volunteers helps ensure quality data collection. The most effective way to support volunteers is to keep them informed about the project and its progress so that they know why they are doing what they are doing and why it is important. Volunteers will devote much time and energy to a project when they feel that they are part of the team. Creating opportunities for local involvement in the study of wetlands resources and their management sends a clear message to the volunteers that public agencies are accessible and local involvement is important. Although they have less specialized knowledge and skill than their professional counterparts, volunteers feel just as strongly about protecting the environment. Many citizens believe that natural resources are truly a part of the public trust and as opportunities open up for cooperative ventures, they are ready to work to protect resource values for all to share. The Trail's End project combined volunteers and scientists in an environmental partnership, and both groups said, "It worked great!"

Volunteers should receive feedback early in the project on how they are doing. Immediate feedback about problems in data recording, questions on technique, or concerns about procedures helps relieve anxiety about data collection problems and helps prevent the possibility of lost data caused by improper use of procedures. During training, it should be emphasized that volunteers are making important contributions to the study and that they are free to telephone the team leaders about any concerns.

If possible, volunteers should be given updates on the data they are collecting, including any preliminary trends, patterns, or new information. Although resource scientists usually wait until data collection is complete to conduct analyses, volunteers appreciate preliminary observations, even though limited, so that they can see the results of their work. Volunteers should not be isolated after they have started work on a particular data collection task. They need to contact each other and share information on experiences and topics of mutual interest. The Local Team Leader can play an important role in maintaining this communication by sending brief newsletters or notes to the volunteers. Any publicity about the project (newspaper, television, newsletters, etc.) should be brought to the attention of the volunteers before it becomes general public knowledge. Weekly meetings during peak data collection times can also keep everyone informed. A debriefing session midway through the project and again near the end help meet this need. At the close of the project, volunteers need to know how project results will be used and why they are important in a larger context. The Off-Site Team Leader and the Principal Investigator can help keep volunteers informed by sending them preprints of journal articles resulting from the project.

Some unstructured social time is important. Volunteers need to have some free time with each other and with resource personnel. Lunches and planned break times give volunteers opportunities for extended conversations with staff on topics of common interest, in this case, wetlands. One major social event, such as a picnic away from the project, with family members included, can add to the overall experience of everyone involved in the project. For the Trail's End Project, a salmon barbecue at the home of the On-site Team Leader provided an excellent setting for informal sharing. The research sponsor needs to accomplish the project tasks, meet deadlines, consider costs, and maintain the highest level of scientific standards, but personal exchange and a shared awareness of the importance of natural systems at work do not have to be sacrificed in the process.

Opening research projects to local volunteers not only aids in data collection and other tasks, it heightens awareness in the community and among resource staff that protecting resources is everyone's job. It is important to remember that many people interested in becoming volunteers are working on other resource projects that may be of interest to the scientific staff. Their participation can broaden the sensitivity of the resource specialists to the interrelated nature of community-based projects.

Benefits for Volunteers and Resource Managers

Local citizens often believe that some distant person is in charge of their resources. Typically, resource managers come to town, do what they do, and leave. There may be some information in the local newspaper about their work, but more likely there will not be unless major issues arise, such as health, natural resource conservation, contested development, or infringement on local government. It is unfortunate that resource monitoring and protection often remain unnoticed. Inviting interested local citizens to become part of a study helps establish a line of communication with resource organizations. It gives interested people an opportunity to contribute to the management process. Most of all, it gives the volunteers a chance to join in the excitement of conducting research and seeing the results used for making better resource management decisions. In the case of Trail's End, it was satisfying and rewarding when the data the volunteers collected passed the EPA review process. The scientists and volunteers had gained open communication, quality data, and ongoing community involvement in the protection of local wetlands.

Involving local citizens at the research level creates a small, but informed group of citizens who are sensitive to the protection of natural resources at a more sophisticated level than the general public. Even though the project ends, and the resource professionals go on to other problems, the new knowledge and awareness, in this case about wetlands and natural history, stays in the community with the volunteers. Local volunteers who helped with the Trail's End project have continued to collect data using the techniques learned during the project. They also added a bird banding element to the study because they wanted to know more about the migratory birds using the wetlands.

Individuals who become involved in a project of this type will begin to influence their friends, coworkers, students, and local governments.

According to their own evaluations made at the close of the project, the individuals who participated in the Trail's End study found the experience and the learning more than worth the effort. The volunteers were doing what they like to do best: being outdoors, protecting natural resources, and sharing that experience with others of similar interests. Volunteers left the project feeling that they had made a positive contribution, enjoyed the opportunity to work with dedicated researchers, become part of a local team, and contributed to wetlands protection. Many also renewed their connection with the earth. As one Trail's End volunteer put it, "The wetland became my friend. I kept going back to say 'hi' to the yellowlegs".

Discussion for The Fight To Save Our Wetlands

1. Propose a set of guidelines for the determination of a wetland. What sort of ecosystem, with what flora, fauna and geological features should it have?

2. How can we settle disputes when the cause was established one hundred years previously? Should the errors of our past ways of planning be given special consideration and perhaps exempted from our new laws?

3. Be the devils advocate and propose a set of rules specific to Alaska. Assume that the Alaskan wetlands are of no concern to the southern states. How would you argue for this position?

4. Do you think the oil industry has Alaska wetlands health in mind? Should there be tougher laws on money making organizations, such as the oil industry, than upon a community, state or government entity?

5. Why is the work of Neal Maine a success? Try and construct the type of organization that he organizes and how he would go about solving the problem.

Introduction

Gillnet Fishing on the West Coast

This issue examines the attempt to protect our marine resources. The sports fishing industry has been vocal in trying to get attention to the fact that our fish are disappearing. Tuna cannot be found where they were once plentiful. The Gillnet initiative is a controversial means of solving this problem...or will it?

The Magnuson Fisheries Conservation and Management Act of 1977 was reapproved in October 1990. This Act bans any vessel in U.S. waters and all U.S. boats from using driftnets longer than 1.5 miles. Trade embargoes are authorized against any country that does not abide by these regulations or does not follow fish limit guidelines in all U.S. waters. Based on 1976 law, the marine fisheries are the property of the American Public to the 200 mile limit. Virtual Population Analysis, VPA's, determine the Magnuson Act quotas known as the infamous 602 guidelines. These guidelines are determined from a series of computer extrapolations that indicate the relative number of fish surviving to spawn the next season. The Japanese squid fishermen are known to use 160 net sections each 100 feet long as one three mile long driftnet. Eight of these nets are fished by each boat each night! This is a 24 mile impenetrable barrier strung across the Pacific Ocean. Squid are an uncontrolled species in International waters so no law is being broken. The United Nations is considering a moratorium on all driftnets if clean targeting cannot be proven. Clean targeting means only the designated species is to be captured and no significant bycatch or any endangered species is taken. In the case of the Japanese squid fishermen, one three mile section of net can bring in 3800 squid each weighing about 5 pounds. Ship board observers have documented the accidental killing of mammals, birds and non-targeted fish in large numbers. Observers on 4% of the Japanese squid fleet noted 59,060 albacore, 58,000 blue sharks, 9,173 sea birds and 914 dolphins. Are the coastal driftnets as deadly and indiscriminate? Is the cost of the ban worth it? Consider the fact that the Fishing Industry in 1990 took about $3.6 billion worth of fish and 64 species were overfished. No royalties were paid for this use of a public resource, but many taxes, fees, licenses and boat loans were obtained from the fishing industry. Commercial fishing is exempt from fuel taxes (a $75-112.5 million subsidy) and are not charged for the Coast Guard or Marine Services such as the weather information and sea temperatures. Vessel loans are guaranteed by the federal government and it is estimated that the federal cost on such programs is about $126 million.

The California Marine Resources Protection Act of 1990, known as Proposition 132, passed with a 55.5% of the vote. This amendment of the State Constitution bans gillnets from within 3 miles of the coast, bans rockfish gillnetting, compensates the gillnetters and establishes open water reserves. The State had 450 licensed gillnetters. By restricting gillnetting from within three miles of the coast the Act will wipe out the industry. All targeted fish are found in the coastal productive zone which extends out to the drop off and in most cases that is within three miles of the coastline. Gillnetters will be bought out or at least recompensed for their losses as they change fishing styles and the money for this will be derived from a $3.00 surcharge on all California salt water fishing licenses. Is this a fair means of protecting our fishery resources? Are the declining fish stocks a fact or are they an artifact of the weather or computer programs? Will this ban have any effect upon this resource? Marine mammal protection was a pivotal point in this legislation and is discussed in Reading 2. There is no shortage of opinions on this issue and tempers are hot.

Barry Thomas

Questions and Concerns

READING 1

As you read this article consider the following:

1. What are the constraints upon a non-profit research organization when it is involved in a public dispute?
2. Do you feel that the data presented was done so in a fair manner and could you now render a fair verdict if you were the judge?

Reading 1

Scientists as Advocates: The Point Reyes Bird Observatory and Gill Netting in Central California

JAMES E. SALZMAN
68 Francis Avenue
Cambridge, MA 02138, U.S.A.

Abstract: *Scientists and scientific organizations can play singularly important roles as advocates in the public resolution of conservation disputes. One scientific organization, the Point Reyes Bird Observatory, was instrumental in resolving an environmental controversy in the early 1980s over the use of gill nets off central California. The gill-net fishery was killing thousands of birds annually. The Point Reyes Bird Observatory's use of its scientific data facilitated negotiation and formed the basis for the threat of litigation. These efforts spurred enforcement of federal wildlife legislation and helped forge a permanent solution acceptable to state and federal environmental agencies, conservation groups, and local fishermen. But there are constraints on the active involvement of scientists in public disputes. The contentious nature of the decision-making forum, the exigencies and uncertainty inherent in policy decisions, and the bias popularly associated with lobbying efforts all tend to dissuade scientists from participation as advocates in public controversies. The gill-net case suggests that when coupled with supportive legislation, focused advocacy—the presentation of relevant data and insistence that it be interpreted accurately and acted upon—is an effective method of achieving biologically sound policies.*

Resumen: *Los científicos y las organizaciones científicas pueden jugar papeles singularmente importantes como defensores en la resolución pública de desacuerdos en conservación. Un organismo científico, el "Point Reyes Bird Observatory," fue crucial en la resolución de la controversia ambiental en los comienzos de los ochenta sobre el uso de redes de acarreo en las costas centrales de California. La pesca por redes de acarreo estaba matando miles de aves anualmente. El uso de datos científicos especificos por parte de "Point Reyes Bird Observatory" facilitó las negociaciones y formó la base para la amenaza de litigio. Estos esfuerzos estimularon la aplicación de la legislación federal en vida silvestre y ayudó a forjar una solución permanente y aceptable tanto para las agencias medio ambientales federales y estatales, como para grupos conservacionístas y pescadores locales. Pero, existen reatricciones en la participación activa de los científicos en disputas públicas. La naturaleza contenciosa del foro para la toma de decisiones; las exigencias y la inseguridad inherente en decisiones políticas y los prejuicios asociados con los esfuerzos de influenciar legislación, tienden a persuadir a los científicos de no participar en controversias públicas. Este caso de las redes de acarreo sugiere que el apoyo focalizado; la presentación de datos revelantes y la insistencia de que sean interpretados con precisión y ejecutados; al emparejarse con legislación de apoyo, es un método efectivo de lograr políticas biologicamente aceptables.*

Introduction

The last two decades have witnessed a tremendous growth of federal environmental legislation. Laws such as the National Environmental Policy Act of 1969, the Clean Air Act of 1970, the Marine Mammal Protection Act of 1972, and the Endangered Species Act of 1973 have come of age, and administrative and statutory frameworks now exist specifically to address environmental policies and disputes. These laws have furthered conservation goals not only because they direct the gov-

Paper submitted June 2, 1988; revised manuscript accepted November 14, 1988.

ernmental mandate to protect the environment, but because they provide private groups the means to shape this process through public comment and litigation. Scientists and scientific organizations can play an influential role in conservation disputes through both research and advocacy. In fact, 21 years ago a group of scientists concerned over the effects of DDT formed the Environmental Defense Fund, now an influential conservation organization. This paper presents a case study of the successful advocacy efforts of the Point Reyes Bird Observatory (PRBO), a research, conservation, and educational organization in northern California, in a recent controversy involving the use of gill nets off central California. The article explores this organization's effective use of scientific data, negotiation, and legislation to forge a permanent solution to a complicated environmental dispute. The concluding section examines the constraints on scientists' participation in public disputes and presents the perspectives of several influential environmental advocates.

A Brief History of the Point Reyes Bird Observatory

In March, 1965, the Point Reyes Bird Observatory was established in the hills of Point Reyes National Seashore just north of San Francisco. The founders, a group of scientists and bird watchers, envisioned a small banding station that would be used to study coastal birds and monitor land-bird migration through the Point Reyes headlands. With sufficient funds, they hoped to support at least one full-time biologist who would work with volunteers. PRBO was in full-time operation by the end of 1965, its residents and volunteers busy trapping, banding, and censusing local avian populations (McCrimmon 1985). PRBO grew rapidly, bolstered by the emergence of the environmental movement.

Neither purely an academic institution nor a research organization, PRBO from its early days recognized its ability to conduct long-term research studies while involving the public through volunteer field work. Its educational activities over the years have included a training program for environmental education teachers, a vigorous volunteer program, and publication of a newsletter. These efforts are consonant with the organization's commitment to raising the public's awareness of environmental issues. In conservation issues, PRBO has exercised influence through its basic and applied research and advocacy of its findings. PRBO's shorebird, seabird, and wetlands research and other activities have given it a respected voice in policy debates. These controversies have ranged from outer continental shelf oil leases, Army Corps of Engineers' dredging near snowy Plover (*Charadrius alexandrinus*) nesting sites, and damage assessments of the biological impact of oil spills, to the impact on bird populations of krill harvesting in the Antarctic. In fact, many of PRBO's long-term studies are directed toward obtaining scientific data to aid conservation and management efforts.

As PRBO has grown, it has retained its character as a singular research establishment. The 1986 budget stood at $860,000, 17 times greater than the budget of 1975. The sources of these funds include: gifts and dues, 30%; contracts, 43%; and grants, 14% (PRBO 1986). PRBO's present staff has 16 full-time biologists, 5 of whom hold doctoral degrees. Including the research associates and volunteers, however, there are more than 50 employees. Throughout its growth, PRBO has remained a scientist's cooperative, promoting conservation when its research efforts can play a constructive role. In the early 1980s, PRBO became directly involved in just such a controversy.

The Emergence of Gill Netting in Central California

Dorothy Hunt lives in the town of Aptos, on Monterey Bay. Dorothy often walks the same stretch of beach near her home to enjoy the fresh air, and to look for dead birds. As a volunteer for PRBO's Beached Bird Project, Dorothy's observations are useful in establishing a data baseline for how many seabirds "normally" die and wash ashore on beaches. On Dorothy's stretch of beach, five or six dead seabirds are a lot to find in one walk. One foggy morning, seven years ago in July, Dorothy found 108 birds washed up among the kelp and seaweed. Frances Bidstrup, another PRBO beach walker working a few miles south of Dorothy, found 185 beached birds on just a half-mile stretch of beach. Frances's and Dorothy's findings were not unique. On that 3-mile stretch of Manresa Beach, over 1,500 birds washed up in one day (Heneman 1981). Gill-net fishing had come to Monterey Bay.

The brightly-colored yellow and green nylon gill nets seem innocuous enough, but they are terribly effective at catching fish and other life that happens to become ensnared in their mesh. Fishermen generally use two types of gill net depending on the kinds of fish they wish to catch. In deep water they use 6-cm-mesh nets, while in shallower water they use a wider 20-cm mesh. Like giant curtains, the monofilament nets are lowered with lead weights to the ocean floor. The float line keeps the 3- to 6-meter-high nets vertical in the water. When large-enough fish hit these invisible threads, only their heads fit through the mesh. Struggling to get free, they tangle their gill covers, are unable to breathe, and suffocate. Two to 24 hours later, the nets are reeled onto large drums and the day's catch is removed. The 20-cm mesh nets work especially well for bottom fish such as Pacific Halibut (*Hippoglossus stenolepsis*), but the nets are indiscriminate, tangling and drowning diving seabirds and marine mammals as well.

Gill nets had actually been in use in Monterey Bay before 1981. In fact, there had been five or six gill-net boats on Monterey Bay throughout the 1970s. In 1981, however, there numbered 25 to 30 gill-net boats, most of them manned by refugees from South Vietnam. While the previous fishermen had used 0.4- to 0.8-km nets, the Vietnamese were setting nets up to 2.4 km long and fishing the same areas much more intensively. Coupled with this dramatic increase in gill-netting activity was an unusually large number of anchovies, which brought the seabirds into shallow waters, now filled with nets. This massive increase in dead seabirds did not go unnoticed. As many as 10,000 seabirds died in July alone, equaling the mortality caused by the major 1971 San Francisco oil spill (Heneman 1981). Spurred on by the Friends of the Sea Otter and other environmental groups, angered beach-goers, local newspapers, and residents of this popular tourist area, meetings began between the California Department of Fish and Game (CDFG), the U.S. Fish and Wildlife Service (USFWS), and private organizations, including PRBO. The parties hoped to find a solution that would maintain the fishery while reducing seabird mortality. As the use of gill nets was unregulated by state law, without enabling legislation CDFG had no power to restrict their use further.

The Search for a Legislative Solution

PRBO's participation in these initial discussions was mainly advisory, using its Beached Bird Project data to show that up to 90% of the seabird mortality was directly attributable to the use of gill nets (Heneman 1981). The main push for legislation, however, came from the Friends of the Sea Otter. In February, State Senator Henry Mello followed the recommendations of CDFG and sponsored bill SB 1475, which restricted gill netting in Monterey Bay to waters deeper than 10 fathoms (18.3 m). This bill also gave CDFG the authority to impose further restrictions if the initial actions were insufficient to reduce seabird mortality to "acceptable levels." As a result of budgetary constraints, however, the bill did not provide funds for CDFG to monitor the effectiveness of the regulations.

The omission of monitoring funds posed a significant problem. Senator Mello's bill presumed that placing the nets outside the shallow waters of Monterey Bay would reduce the drastic seabird mortality, but this was only an assumption. No one knew what level of mortality would occur in waters deeper than 10 fathoms. The Moss Landing Marine Lab had started monitoring the incidental catch by gill-net boats in 1981 to determine mortality (since many dead seabirds discarded from boats do not wash ashore) (Carter, personal communication). Unless the monitoring was continued, an unacceptable number of seabirds might still be killed, but a decline in seabirds found along the beaches would lead to the opposite, and incorrect, conclusion. In recognition of this danger, PRBO, along with the Monterey Peninsula Audubon Society, pledged to raise the $5,000 needed to continue the monitoring program through the summer (Heneman 1982). PRBO also solicited the aid of volunteers to help CDFG in their own beached bird censuses. Up to now, PRBO had played in important advisory and supporting role, but Monterey Bay was 100 miles south of PRBO, and the local organizations had taken the lead. This all changed in the summer of 1982.

The 17 June 1982 edition of the *Point Reyes Light,* a local Marin County newspaper, prominently featured an article titled, "Birds Face Net Peril" (PRL 1982*a*) (see Fig. 1). The paper reported that "in the past 2 years thousands of shore birds have died in gill nets used by fishermen in Monterey Bay. The same fishermen are now leaving that area and coming north seeking new territory. . . . Unfortunately, the fishermen are bringing the identical problems with them to our Bay" (PRL 1982*a*). In early June two Vietnamese boats, the *Cole* and the *Rhonda Lea,* mindful of the pending legislation, had moved up the coast from Monterey Bay and set 3.2

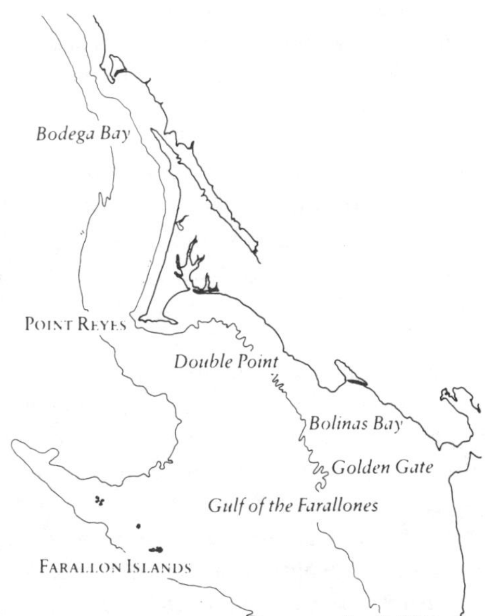

Figure 1. The Central California coast. Monterey Bay is just south of Devil's Slide. Source: *Map by Jay Penniman in Point Reyes Bird Observatory Newsletter, 1986, 75:3.*

km-long nets only 400 meters off Stinson Beach in Marin County (MIJ 1982*a*). The next day, two dead seals, an injured baby harbor porpoise (*Phocoena phocoena*), and 200 dead seabirds were found washed ashore (CP 1982*a*). PRBO's headquarters was only a few miles from that beach, and it quickly became involved.

Burr Heneman, then executive director of PRBO, had previously been involved with the gill-net issue in Monterey. He spoke to the press, was quoted in all the local papers, and wrote an article in the *PRBO Newsletter*. Within a week, Heneman traveled to Sacramento to obtain the consent of Senator Mello to amend his bill to include Marin and Sonoma counties and San Francisco Bay. Heneman met with E. C. Fullerton (director of CDFG), state legislators, and Zeke Grader (president of the Pacific Coast Federation of Fishermen's Associations). Neither Grader nor Fullerton supported a legislative ban on the use of gill nets along the entire coast. Thus a compromise was reached that granted CDFG the authority to prohibit or restrict the use of gill nets temporarily following a public hearing, if it was determined "that mortality in any local population of any species of seabird or marine mammal is occurring at a rate that threatens the viability of the local population, as a direct result of the use of gill nets" (Atkins & Heneman 1987). Governor Brown signed SB 1475 into law on 28 June, 1982.

The next day, a hearing was called by CDFG in Point Reyes to discuss the gill-net issue. By this time, the issue was extremely charged. Local residents had gathered over 700 signatures on petitions calling for the banning of gill nets off Marin County (MIJ 1982*b*). The *Cole* and the *Rhonda Lea* were still working up and down Stinson Beach, but now surrounded by government officials. As the *Coastal Post* reported: "Park lifeguards in their Zodiac vessel went out to the fishing boat and watched them pull in their nets. Although park employees estimate 200 dead California Murres [Common Murre = *Uria aalge*] and 10 or 15 Common Cormorants [Brandt's Cormorant = *Phalacrocorax penicillatus*] were visible in the nets, no mammals were seen. During the net pulling a news helicopter hung over the fishing operation for two hours. Federal and state officials waited on the beach to see if there was any evidence that the nets had caused the deaths and injury" (CP 1982*a*). Although none of the marine mammal deaths could be conclusively linked to gill-net fishing, the public regarded the connection as more than coincidental. The sudden appearance of Great White Sharks (*Carcharodon carcharias*), which led to the closing of the beach for a week in July, was also blamed on the dead fish and seals caused by gill netting (Heneman 1983). Indeed, by the time of the 29 June hearing, at least 13 Harbor Porpoises, fourteen Harbor Seals (*Phoca vitulina*), and a six-foot Blue Shark (*Prionace glauca*) had washed up on the Marin County coast (PRL 1982*b*).

At the hearing, PRBO testified along with other citizen groups, local government officials, and fishermen. But PRBO was the only group with actual data on what was happening. PRBO had observed six net pulls, about 10% of those that actually had occurred, and counted 400 dead seabirds (PRL 1982*b*). This testimony, as in Monterey, effectively dispelled the notion that "there wasn't a problem." As Heneman explained, "PRBO was not perceived as a biased advocate. We were an advocate for the right thing happening to wildlife and basing whatever happens on solid data. No other groups had that" (Heneman, personal communication). Zeke Grader, the fishermen's lobbyist, held a different perspective: "PRBO certainly played to a degree the role of advocates. But we kept their bias apart from their scientific information; and their information played a vital role. The fishermen felt the information was valid because it fit in with what they were seeing out there" (Grader, personal communication). In spite of PRBO's data, local fishermen, who for years had used hand-pulled gill nets (100 to 250 meters), were vocal about their concerns. Stated Josh Churchman, who fishes out of a 17-foot motorboat, "if you force us to move our nets to deeper water, I'm out of business—I'll lose my livelihood" (MIJ 1982*c*). While Churchman and others favored a ban on certain types of net, PRBO favored a ban on all gill netting under ten fathoms, along with continued monitoring. As Heneman stated in a *Coastal Post* editorial, "if you set any kind of nets where there are birds, you will catch birds" (CP 1982*b*). CDFG concurred with PRBO's proposal, adding that "we will not take a broad axe approach and just wipe out gill-net fishing. However, we may have to take a scalpel and go deeper until we find the source of the problem" (MIJ 1982*c*).

At the time of the hearing, seven gill-net boats were operating off Marin County, three of those Vietnamese. Within days of the regulations taking effect, PRBO learned of heavy seabird mortality down south along the San Mateo County coast. CDFG quickly responded with another hearing and new regulations. As Heneman explained, "the gill netters were finding new areas to fish faster than we were getting the locations covered by legislation. Now we were beginning to get reports of seabird and marine mammal kills following the familiar pattern along the Big Sur coast, near Morro Bay, and off Santa Barbara County. What was once a local Monterey problem was now regional in scope" (Heneman 1982). Still visiting Sacramento, Heneman helped draft a bill that would have given CDFG broad authority to regulate certain commercial fishing gear, such as nets that extensively damaged marine life. The bill was killed in the legislature by heavy lobbying from the southern California seafood industry (Heneman 1982). And the temporary ban on gill netting off Marin County expired in September. The monitoring, however, continued. CDFG assigned two department marine biologists to study further the impact of gill netting while PRBO and the International Council for Bird Preservation pledged to obtain the $5,000 necessary to continue the monitoring program (Heneman 1982).

More Temporary Restrictions

By the summer of 1983, many were dismayed by the situation. CDFG's temporary bans had expired the previous September, and in June, 1983, thousands of seabirds were once again washing up on the shores of San Francisco and Marin counties (PRL 1983*a*; CP 1983*a*). At the fourth CDFG hearing on gill netting held in early June, however, CDFG Director Fullerton concluded that the evidence presented did not warrant imposing temporary restrictions (PRL 1983*b*). The CDFG study started the previous year had not yielded any conclusive data.

CDFG's inaction may have seemed arbitrary, but it was consistent with the agency's philosophy of minimizing the impact to the gill-net fishery while protecting populations of species that were caught incidentally. There were still too many unanswered questions. As Heneman explained, "the phrase 'acceptable incidental kill' is the crux of the problem. Obviously, as long as netting continues you can't cut the losses to zero. But to tailor regulations that will reduce bird kills without bankrupting fishermen, you need a lot of hard data. Right now Fish and Game lacks the manpower and funding to do a thorough job of gathering all the information" (NF 1983). What species were being caught in nets made of different materials, with different mesh, set at different depths, at various times of year? A flat ban on all gill netting would make these questions moot, but Fullerton refused to consider such an approach. These questions were still unanswered when, in early August, hundreds more seabirds washed ashore along the San Francisco, San Mateo, and Marin county coasts (SFE 1983). The new head of CDFG, Howard Carper, called for a hearing which proved to be the most emotional gathering yet.

The August hearing began dramatically when the president of the San Francisco Society for the Prevention of Cruelty to Animals proclaimed, "we're talking about thousands and thousands and thousands of birds being slaughtered," while associates dumped a garbage can full of 72 foul-smelling individually wrapped seabirds onto the carpeted floor (MIJ 1983). On the opposing side, Dave Kruegel, the sponsor of a Vietnamese boat, claimed, to a chorus of boos, that a ban on gill nets would be a form of racism against Vietnamese fishermen (MIJ 1983). PRBO and CDFG biologists presented data that showed the continuing effect of gill netting on local seabird and marine mammal populations. The number of gill-net boats had increased from 34 boats in 1981 to 54 in 1982. At the time of the hearing, 74 gill-net boats were operating off central California (NF 1983). Part of this dramatic increase was due to an extremely poor salmon season, which forced salmon boats to fish with gill nets to make ends meet. PRBO concluded that this year's mortality of common murres alone could total 17,000 birds, on the order of 10% of the breeding population (Heneman 1983; MIJ 1983). Other environmental, civic, and sport-fishing groups spoke, and all but two of the gill-net fishermen present acknowledged the need for some form of regulation. Carper responded with temporary gill-net restrictions along the coast effective until 16 October, 1983 (Heneman 1983).

The Effect of Gill Netting on the Common Murre

The PRBO data proved especially useful at the hearing for it both reinforced and supplemented CDFG's findings. It also provided firsthand accounts of fishermen's attempts to conceal the incidental catch by weighting garbage bags full of dead birds or casting them out along rocky, inaccessible shores (Heneman 1983). Most important, though, was PRBO's ability to project the effects of mortality on the viability of the population. Since 1969, PRBO has served as the caretaker of the Farallon Islands, a National Wildlife Refuge that hosts the largest Common Murre breeding colony on the central California coast. In the mid-1800s, the murre colony on the Farallones may have numbered as high as 400,000 (Carter 1984). But by the end of the century the murre population on the Farallones had almost been wiped out by organized companies of egg gatherers, who satisfied the forty-niners' desire for "sunny side up," at a price. Over the years, the egg companies took as many as 12 million murre eggs (TW 1986). In later years, oil spills and bilge flushing outside the Golden Gate kept the population at low numbers. In 1972, three years after PRBO started managing the Farallones Islands, the common murre population stood at just 22,000. By 1982, the population had grown to 88,000 birds (Carter 1984). But over 90% of the seabird tangled in gill nets were common murres (NF 1983). Drawing on their data, PRBO scientists pointed out that such high levels of gill-net mortality would result in a rapid decline in the breeding population.

All the environmental groups could point out that birds were dying, and PRBO and CDFG could state how many, but only PRBO could show the long-term consequences on a specific population. As the *National Fisherman* reported, "Heneman's testimony provided the only concrete appraisal of the impact of net fishing on the overall bird population" (NF 1983). Don Schultz, a CDFG biologist, concurred: "PRBO was the public group with the most pertinent data regarding the status of the seabirds. In fact, they had more detailed information than the agency. When an outside entity provides data for far-reaching decisions, they are open to criticism; but the validity of PRBO's data was never challenged. They were recognized as experts in their field and the data was collected in a scientific manner. They were an integral part of the discussions" (Schultz, personal communication).

PRBO was convinced that the murre population could not sustain such a high level of mortality. Yet the pattern of a massive kill followed by a CDFG hearing and a temporary closure seemed to have become an endless loop. Five separate hearings had not stopped the mortality, and, on the fishermen's side, the stopgap measures simply pushed the incidental-kill problem around to other areas, where more boats competed for the same limited resource. Through the winter of 1983–1984, Heneman and Grader drafted a bill that provided for year-round and seasonal closures in particularly sensitive areas, such as that around the Farallon Islands. The bill also restricted the permissible length and height of nets and placed a one-year moratorium on the issuance of gill net permits. It did not, however, address marine mammal mortalities. The bill, SB 2266, was supported by all of the interested parties, including CDFG, California Gillnetters Association, National Audubon Society, and Marin County, and was signed into law on 20 June, 1984, by Governor Deukmajian (Heneman 1984).

While Heneman and Grader both told the press that "this is not a total solution" (MIJ 1984), the passage of the 1984 bill seemed, for the moment at least, to be an adequate resolution of the gill-net controversy. CDFG and PRBO monitoring studies continued, and, indeed, fewer numbers of murres were dying than in previous years. The question, though, was whether this represented a real decrease in mortality, or whether simply fewer birds were left to be killed. By early 1985, PRBO was convinced that the gill-netting situation was critical. In 1983, a cyclic global climatic event, called El Nino, had caused breeding failures and reduced seabird populations, so gill-net deaths had to be calculated from an already depleted population (PRL 1983*c*). Even taking the effects of El Nino on the murre colony into account, however, PRBO biologists found that gill netting still annually killed in the order of 10% of the murre population (Carter & Ainley 1986). Disseminating this data to other environmental groups, PRBO convinced them of the need for still stricter gill-net controls. But the commercial fishermen were unwilling to negotiate, believing the problem had been adequately resolved, and CDFG wanted to wait another season or two before evaluating the closures' adequacy. Ironically, while the closures had not halted the offshore mortality, they had been effective in eliminating the massive numbers of dead birds on beaches, thus removing the issue from the public's eye and denying the environmental groups their most effective tools: an outraged media and public. Frustrated at the state level, environmental organizations looked for the first time to the federal government.

The Initiative in Washington

Up until now, there had been little or no federal involvement in the gill-net affair beyond funding some of the monitoring programs. This seems odd, because the incidental kill of seabirds and marine mammals suggests conflict with the Migratory Bird Treaty Act, the Marine Mammal Protection Act, and the Endangered Species Act, all federal laws. The cause for the federal inaction lay in an informal agreement between USFWS, NMFS, and CDFG. As Heneman explained later, "unofficially, the position of the management staff at these [federal] agencies was that since the fisheries were state-managed and conducted in state waters, the solution was up to the state" (Atkins & Heneman 1987). The enforcement staff of the federal agencies did not want to get involved, either, because of the difficulty in obtaining incidental take convictions. At the same time, while California state law does incorporate the federal wildlife laws and CDFG did have the authority to enforce the laws, CDFG was understaffed and hesitant to step in and enforce provisions that the federal agencies were clearly avoiding and that contradicted their own stated policy of maintaining a viable gill-net fishery (Atkins & Heneman 1987). From the beginning, environmental groups had been content with federal agencies standing on the sidelines while CDFG tried various regulatory solutions. And the regulations had been effective to some extent, for the incidental take had been reduced from 1982 levels. But it was clear to PRBO that the problem had not gone away, and it convinced a number of national groups to take the struggle to Washington.

In the spring of 1985, Congresswoman Barbara Boxer and four other California legislators officially requested the USFWS and NMFS to carry out their obligations under federal law and act against illegal incidental take of seabirds and marine mammals by gill netting (CP 1985). They demanded a reply within sixty days. Under the Migratory Bird Treaty Act, 16 U.S.C. §703 et seq., it is illegal to kill birds included in the treaty unless the Secretary of the Interior adopts excepting legislation. The Common Murre is listed in the treaty. The Marine Mammal Protection Act, 16 U.S.C. §1531 et seq., similarly prohibits the killing of marine mammals subject to certain exceptions provided by the Secretary of Commerce. The Endangered Species Act, 16 U.S.C. §1531, prohibits the killing of species listed as "endangered" or "threatened," except pursuant to a research permit. The Southern Sea Otter (***Enhydra lutris nereis***) is listed as a threatened species. Thus, it appeared that the gill-net incidental kills were in blatant violation of federal and state law. USFWS's response acknowledged that the deaths had occurred and that proper preliminary steps might not have been taken. NMFS simply stated that they were presently working on the issue (Heneman, personal communication). Thus both agencies admitted that a problem existed, but neither committed itself to act. The initiative in Washington had apparently failed.

Up to now, PRBO had based its population data on the Farallon Island colonies. While USFWS's response in Washington had been noncommittal, the agency recognized that a problem existed and approached PRBO about the situation. Two PRBO biologists, David Ainley and Harry Carter, had previously studied the impact of gill netting in Alaska and British Columbia (Ainley et al. 1981; Carter & Sealy 1984). PRBO proposed that aerial censuses of murre colonies up and down the coast would provide exact and irrefutable evidence of the population's decline (Carter, personal communication). USFWS concurred, and beginning in the summer of 1985, USFWS and PRBO biologists took aerial photographs and conducted ground counts of murre colonies along the California coast. Meanwhile, Ainley continued an extensive study of the distribution of murres in the Gulf of the Farallones. This data would be helpful in identifying locations for gill netting where few seabirds would be affected. Heneman, now a consultant for the Marine Mammal Commission, was still working on forging a political solution. A workshop in January of 1986 succeeded in bringing together all the interested parties, including the federal agencies and the Vietnamese Fishermen's Association, but little was accomplished beyond heated debate. To many parties, gill netting was a resolved issue.

New Data and Threat of Litigation Break the Stalemate

In the summer of 1986, however, the aerial census data provided a shocking rebuttal. Flying over Devil's Slide Rock, USFWS and PRBO biologists expected to see the 2,300 murres that had bred there as recently as 1982. But there were no murres. The entire colony had been wiped out in only four years (Carter & Ainley 1986). As PRBO and USFWS biologists pored over the 1985 and 1986 aerial photographs, counting actual numbers of murres in each colony, they found that while murre colonies in northern California had remained stable since 1982, many colonies in central California had declined by up to 80%, even faster than the baseline colony monitored on the Farallones. Most of this decline was directly attributable to gill-netting (Takekawa et al. 1988). CDFG data also showed a harbor porpoise mortality rate of 200 to 300 per year out of a population as low as 3,000 porpoises (Heneman, personal communication). PRBO passed this new information to all the gill-net parties, issued a press release, and published a comprehensive article in its newsletter (Carter & Ainley 1986). The pace of events quickened.

The Washington-based Center for Environmental Education and the National Audubon Society, both of whom had previously watched from the sidelines, agreed to lobby USFWS and NMFS. Realizing that the threat of litigation was now the most effective alternative to break the political stalemate, Heneman encouraged a number of organizations, such as the United Anglers of California and Marin Peninsula Audubon Society, to raise funds to evaluate the strength of the gill-net case. The 17-page legal opinion by a Sacramento law firm concluded that the incidental gill-net mortalities of murres, Harbor Porpoises, Harbor Seals, and Southern Sea Otters were all violations of federal and state law. The attorneys declared that "the current problem does not seem to stem from any failure in law, but rather stems from a failure by those charged with the implementation of the law to fully carry out its mandate." In other words, the environmental groups had a very strong case and, thus, a new lever to force the government into action.

Encouraged by the strength of its legal position, later in the spring of 1986, the Center for Environmental Education and the National Audubon Society petitioned the Secretaries of Interior and Commerce (who oversee USFWS and NMFS, respectively) to enforce the Marine Mammal Protection Act and the Migratory Bird Treaty Act (Heneman 1986). In California, the Marin Audubon Society, the Whale Center, and the United Anglers of California petitioned the state attorney general to provide CDFG with a legal opinion of its responsibilities toward gill-net incidental mortality under state law. The attorney general's office issued its opinion to CDFG, stating that there was an "unequivocal prohibition against taking [of seabirds and harbor porpoises] under State and Federal law." Spurred by this strong opinion and concern from Washington and Sacramento, representatives from CDFG, USFWS, and NMFS met with the Center for Environmental Education lobbyists, Heneman, and PRBO biologists Ainley and Carter to discuss possible remedies. For the first time, the federal agencies conceded that serious action needed to be taken in order to protect the wildlife (Carter, personal communication). By September 1986, these meetings had expanded to include interested environmental groups and sport and commercial fishing interests: the same groups, in fact, who had reached an impasse nine months earlier.

No agreements were forthcoming. While the environmental groups had an undeniably strong legal case to force greater governmental enforcement efforts, the length and costs of litigation were daunting, as were the problems of enforcing a court order. For a long-term solution, the environmental groups needed final and effective legislation. But the backing of the fishermen was necessary for passage of any legislation, or their lobby would oppose it. In November, CDFG revealed its proposed regulation: closure of all fishing areas inside twenty fathoms. Although more comprehensive than the 1984 bill, the CDFG proposal angered both sides. The environmental organizations withheld support because PRBO's data on the distribution of murres indicated that gill-net fishing beyond twenty fathoms might actually increase mortality. The regulations were no benefit to the fishermen, either, for they removed the best gill-net fishing areas and made the White Croaker (*Gentonemus lineatus*) fishery, dominated by the Vietnamese, economically impractical. CDFG stood firm, however, and refused to consider lesser restrictions while likewise refusing to shut down the gill-net fishery (Carter, personal communication).

As Heneman, Grader, and others negotiated behind closed doors, another approach emerged: shifting some gill-net fishermen into a new fishery with alternative fishing methods (such as seine or traps). This proposal allowed CDFG to impose strict restrictions and permanent closures in the Gulf of the Farallones and other areas of high mortality. In exchange for the support of the Pacific Coast Federation of Fishermen's Associations and the Vietnamese Fishermen's Association of America, the compromise provided low-interest loans to fishermen who switched from gill nets to other equipment. In fact, the Vietnamese Fisherman's Association began to explore switching gear at this time with seed funding provided by fishing and environmental groups, including PRBO. The state, eager to bring the conflict to an end, did not oppose the funding provisions of the bill. With a long-term resolution of the problem in hand, the environmental groups agreed to a one-year period for introduction of the new regulations. Governor Deukmajian signed the bill into law in September 1987.

Today, the incidental take of seabirds and marine mammals from gill nets in central California is a resolved issue. A Nearshore Fisheries Research Advisory Committee has been formed from the various groups involved in the issue, including PRBO scientists. This committee is charged to evaluate the economic and biological consequences of alternative gear and methods, such as nighttime fishing, trolling, and trawling.

PRBO's Influence

Throughout this controversy, PRBO played a crucial role. And, surprisingly, neither government, fishermen, nor conservation groups viewed PRBO as a biased antagonist. Yet, as Heneman described later, at various times in the dispute, PRBO "provided its own gill-net impact data, helped collect and interpret agency data, pushed CDFG to conduct an adequate monitoring program, raised private funds to supplement the CDFG monitoring budget, negotiated regulations that were incorporated into several bills (all of which were lobbied through the legislature and across the governor's desk), pressured the federal agencies to become involved, assisted fishermen in switching to alternative gear, and kept PRBO members and the public informed through newsletters and the news media" (Atkins & Heneman 1987; Heneman, personal communication). In fact, PRBO and Heneman were mentioned by name in over 70 articles in newspapers and magazines. Central to PRBO's ability to maintain an objective position in the eyes of the participants was its role as a scientific research organization.

From the beginning, PRBO chose to restrict its contributions to those based on scientific data. As Heneman puts it, "we went down a road, right or wrong, of trying to nail down the fact that gill netting was having a biological impact." While this approach may seem indirect, it suited the situation well because CDFG had competing obligations: protection of wildlife and preservation of the fishing industry. The approach of environmental groups, translating support from the public into political action, was effective only to a point, Carter explained, because "the fishing industry also has strong emotional and political support and CDFG did not want to shut down the gill-net fishery." If CDFG was going to act incrementally, on what basis could its actions be challenged?

PRBO's use of its data in both public and private contexts was critical in preventing a stalemate and in insuring the adequacy of CDFG regulations. As Heneman reflected: "Information is power. You're in a terrible situation going into a room facing a bunch of fishermen with wives and children. 'You're killing birds.' 'No we're not.' Then what? You can't just develop the information, put it on paper and launch it into the blue. Nothing will happen. You must take the information, pursue it, and put the pressure on. You have a responsibility to find someone who can do something with it." Zeke Grader concurs: "The key to PRBO's effectiveness was the presentation of their information. They were very careful, very objective. They didn't say, 'it's your fault, fishermen.' They laid it out and said, 'here's the effect from oil spills, from gill netting, etc.' It came from the perspective of scientists who were actually out there, not groups who heard about it and came in later to say it was wrong. In all my dealings, that's pretty rare."

Basic research, such as investigating species' behavior and sources of mortality, is essential to improve our understanding of how to manage most effectively for biological diversity and viable populations. This research also aids management agencies and advocacy groups in identifying areas where their intervention will be biologically most significant. Yet, as the gill-net case study illustrates, research alone (i.e., simply reporting data) is often too faint a voice to be heard amid the din of controversial policy debates. PRBO did not simply report its findings and disappear, leaving the "policy makers" to resolve the issue. Because PRBO knew more about the biological impact of gill nets than anyone else, it remained an active participant to insure that informed judgments were made. Ainley's and Carter's previous experiences with gill netting proved important in providing irrefutable evidence of a population decline and in constructing scientifically sound solutions to reduce mortality. The volunteers at PRBO also were indispensable in collecting the sheer volume of data needed to make reliable conclusions. Heneman served a number of roles. He not only acted as the voice for PRBO biologists, but he also served as a fulcrum, facilitating movement and compromise from both environmentalists and fishermen.

In many ways, PRBO was an adjunct of the more vocal and visible environmental groups. While dumping a trash barrel of dead birds may have impressed the media, providing indisputable evidence of mortality at the population level was far more useful to CDFG. Both actions were important to the final outcome. But PRBO's data and scientific expertise might not have been accepted without question had it been viewed as part and parcel of advocacy. As Heneman and Carter contend, "being evenhanded and fair, being seen as scientists, is the main, precious thing that an organization like PRBO has to protect in a public policy issue. That's why it's inappropriate to go first to lawyers. That may be the right thing to do in that situation, and you can privately urge other groups. But PRBO can't do it." In fact, when, in 1986, the environmental groups were considering litigation, PRBO chose not to be included as a party, even though its data were the basis of the suit (Carter, personal communication).

Scientists and Focused Advocacy

There exists an ambiguity over the role of scientists as advocates. Judging from the gill-net case study and the nature of public decision making, advocacy seems far more effective in influencing policy than education or research alone. And surely, PRBO, despite its protestations, was an advocate at least on behalf of viable species populations. Yet it is at the mere recognition of advocacy that scientists and scientific organizations often balk. Dr. David Wilcove, senior ecologist with the Wilderness Society, acknowledges that "there exists the potential for far more cooperation between the advocacy groups and the scientific community than we enjoy at the present. Environmental disputes are becoming increasingly more technical in nature, and the input of the scientific community is definitely needed. But the nature of advocacy battles does not appeal to a lot of scientists" (Wilcove, personal communication). The contentious forum for resolving conservation disputes does seem to discourage many scientists from becoming involved. As Harvard Law School's Professor Richard Stewart, a prominent commentator on environmental law, explains, "scientists don't like the framework for resolving issues. They feel that the law oversimplifies and distorts science, and that the decisions are often made by scientific ignoramuses" (Stewart, personal communication). On the other side, advocates are often frustrated by the complexity of scientific issues. They want the scientists to provide strong, unambiguous statements that will sway a jury or policy maker (Black 1988).

Michael Bean, an attorney with the Environmental Defense Fund (EDF) and an authority on wildlife law, acknowledges the tensions inherent in scientists' involvement with advocacy groups. "Policy-makers and advocacy groups work in a milieu which requires making decisions every day, often without necessary knowledge. They must draw the best inferences they can with the available information. But some scientists require a higher degree of certainty about cause and effect" (Bean, personal communication). Doug Foy, director of the Boston-based Conservation Law Foundation, has reached the same conclusion: "Many scientists would rather be silent than give their best informed opinion, even if it is scientifically valid. The problem is not the absence of time; it's the refusal to be heard. Advocacy groups need scientists to stand up and be counted. We need their representation to offset the flaks" (Foy, personal communication). Scientists may well feel uncomfortable with the exigencies of decision making, and the uncertainties inherent in the process. As Wilcove describes, "when the expert witness begins to qualify everything he or she says, the advocate loses patience. Scientists must learn to convey their knowledge with sufficient clarity and persuasiveness to win in the advocacy arena."

To some extent, it appears that scientists view advocacy as antithetical to objectivity. Bean relates that "a number of scientists interested in joining EDF have expressed concern that they are burning their bridges and can't go back into academia and pure science, that they are tainted by joining a pure advocacy group. They believe it will interfere with their future opportunities." Scientists at PRBO also voiced concerns that an organization's participation in public disputes, if not carefully thought out, might have repercussions both in funding and in the esteem of other scientists. As Heneman explained, "PRBO wants to be the kind of organization that isn't perceived as a Sierra Club; PRBO is a science organization with deeply held convictions about a few issues it knows a lot about" (Heneman, personal communication). Heneman's distinction is critical, for the gill-net case study indicates that PRBO exercised influence through the parties' complete acceptance of its data, not through the mere fact that biologists were advocating a position.

More precisely, the mandate of government environmental agencies stems from legislation, and laws such as the Endangered Species Act and the Marine Mammal Protection Act are founded upon scientific definitions such as critical habitat and viable population. In any specific dispute, giving meaning to these legal terms requires data. Thus the value of PRBO's research prior to the advent of the gill-net controversy cannot be overestimated. The beached-bird counts, the murre colony data, and the harbor seal studies were all invaluable for understanding and quantifying the mortality caused by gill nets. Moreover, this data was the driving force behind the negotiated settlements. Every time legislation was passed, the bills had either the support or the neutrality of the fishing lobby. Each bill was written with fishing-group lobbyists, and each went a step farther toward a final solution. But negotiation alone proved insufficient to halt the gill-net mortality. The threat of litigation was crucial to resolving the conflict. Without the power of federal wildlife legislation, the 1984 bill would probably still be the law, coupled with new cycles of seasonal closures. Only when the federal agencies faced the likelihood of fighting, and losing, a lawsuit, did they pressure CDFG to reach a permanent solution.

PRBO's data were thus crucial, but advocacy organizations can rarely afford to pay researchers to count dead birds on the off chance that the data might prove useful at some future date. As Stewart notes, "it is fairly rare to have a scientific institution focused on a particular set of issues congruent with a conservation dispute." Yet this fortunate congruence assured PRBO access to the decision makers.

PRBO's involvement and influence in the gill-net controversy shows how scientists can be extremely effective in policy disputes and maintain their objectivity if they employ a focused advocacy: reporting data and pressing to insure that the information is interpreted correctly and acted upon. Central to focused advocacy is basic conservation-oriented research. In retrospect, PRBO was both smart and lucky. By intentionally directing its research efforts toward biologically sensitive areas and species, PRBO was able to produce data that addressed biological questions posed by gill netting. Litigation can take years, but basic research often takes more years. Moreover, there is no way prior to a conservation crisis to insure that scientists will have performed relevant basic research. The probability of confluence can be increased, however, by supporting research efforts directed toward specific environmentally sensitive areas.

It remains an open question whether a lawsuit would have resulted in a preferable outcome for the fishermen and environmentalists. In retrospect, negotiation did forge a long-term solution acceptable to all parties, a result that a court-ordered settlement probably would not have achieved. Negotiation may also have been preferable for the wildlife, for gill-net mortality is not restricted to California. Each year hundreds of thousands of seabirds are killed in the North Pacific and North Atlantic fisheries (Ainley et al. 1981; Piatt et al. 1984; Piatt & Reddin 1984). Because there are a number of nations involved, such as Japan, Greenland, and Denmark, legal obligations are indistinct and difficult to enforce over such large ocean areas. These conservation problems differ substantially from those present in California, where a small number of fishermen worked in a discrete area and were subject solely to American law. Nonetheless, the resolution of the gill-net controversy off central California does offer a model and hope for future multinational agreement.

Acknowledgments

This article was written under the Winter Term Writing Program at the Harvard Law School. The author is especially grateful to Harry Carter for his extensive editorial and scientific suggestions. Burr Heneman, David Wilcove, Martha Minow, and Eric Fajer also critically reviewed the manuscript and provided many helpful comments. The author thanks Richard Stewart, William Bossert, Michael Bean, Doug Foy, and Ira Rubinoff for sharing their thoughts on science and advocacy. The Point Reyes Bird Observatory was most generous in providing the use of its facilities, resources, and photographs.

Literature Cited

Ainley, D. G., A. R. DeGange, L. L. Jones, and R. J. Beach. 1981. Mortality of seabirds in high-seas salmon gill nets. Fisheries Bulletin 79:800–806.

Atkins, N., and B. Heneman. 1987. The dangers of gill netting to seabirds. American Birds 41:1395–1403.

Black, B. 1988. Evolving legal standards for the admissibility of scientific evidence. Science 239:1508–1512.

Carter, H. R. 1984. Rise and fall of the Farallon common murre. PRBO Newsletter 72:1–3,11.

Carter, H. R., and D.G. Ainley. 1986. Disappearing murres. PRBO Newsletter 75:4

Carter, H. R., and S. G. Sealy. 1984. Marbled murrelet mortality due to gill-net fishing in Barkley Sound, British Columbia. Pages 212-220 in D. N. Nettleship, G. A. Sanger, and P. F. Springer, editors. Marine birds: their feeding ecology and commercial fisheries relationships. Canadian Wildlife Service Special Publication.

Coastal Post. 1982*a*. Vietnamese fishing nets may be responsible for two dead seals and injured porpoise pup. 14 June, 1982. Newspaper article, Marin County, California.

Coastal Post. 1982*b*. 6 July, 1982. Editorial, Marin County, California.

Coastal Post. 1983*a*. Off-limits gill netting area asked. 6 June, 1983. Newspaper article, Marin County, California.

Coastal Post. 1985. Feds asked to help with illegal gillnetting. 4 March, 1985. Newspaper article, Marin County, California.

Heneman, B. 1981. Why count dead seabirds? PRBO Newsletter 55:2.

Heneman, B. 1982. Monterey Bay gill net legislation — help needed. PRBO Newsletter 58:7.

Heneman, B. 1983. Gill nets and seabirds 1983. PRBO Newsletter 63:1, 15, 10.

Heneman, B. 1984. Gill net news. PRBO Newsletter 65:7.

Heneman, B. 1986. Gill net news. PRBO Newsletter 75:1–3, 10.

Marin Independent Journal. 1982*a*, New law cracks down on gill net fishing. 1 July, 1982. Newspaper article, Marin County, California.

Marin Independent Journal. 1982*b*. Bill limiting gill net fishing faces Point Reyes hearing. 23 June, 1982. Newspaper article, Marin County, California.

Marin Independent Journal. 1982*c*. Fishermen fear new gill net law. 29 June, 1982. Newspaper article, Marin County, California.

Marin Independent Journal. 1983. Gill-netting debate renewed. 1 August, 1983. Newspaper article, Marin County, California.

Marin Independent Journal. 1984. Fishermen, environmentalists hail gill-net law. 22 June, 1984. Newspaper article, Marin County, California.

McCrimmon, D. A. Quo vadis? 1985. PRBO begins its second 20 years. PRBO Newsletter 69:1, 6.

National Fisherman. 1983. Fish and Game orders more closures for California gillnetters. November.

Piatt, J. F., D. N. Nettleship, and W. Threlfall. 1984. Net mortality of common murres and Atlantic puffins in Newfoundland, 1951-81. Pages 196-206 in D. N. Nettleship, G. A. Sanger, and P. F. Springer, editors. Marine birds: their feeding ecology and commercial fisheries relationships. Canadian Wildlife Service Special Publication.

Piatt, J. F., and D. G. Reddin. 1984. Recent trends in the west Greenland salmon fishery, and implications for thick-billed murres. Pages 208-210 in D. N. Nettleship, G. A. Sanger, and P. F. Springer, editors. Marine birds: their feeding ecology and commercial fisheries relationships. Canadian Wildlife Service Special Publication.

Point Reyes Bird Observatory. 1986. Annual report, 1986. Point Reyes Bird Observatory, Point Reyes, California.

Point Reyes Light. 1982*a*. Birds face net peril. 17 June, 1982. Newspaper article, Point Reyes, California.

Point Reyes Light. 1982*b*. State can't control gill nets. 1 July, 1982. Newspaper, Point Reyes, California.

Point Reyes Light. 1983*a*. More birds killed by gill nets. 1 July, 1982. Newspaper article, Point Reyes, California.

Point Reyes Light. 1983*b*. Fish and Game refuses to regulate gill nets. 30 June, 1983. Newspaper article, Point Reyes, California.

Point Reyes Light. 1983*c*. Gill nets. 4 August, 1983. Newspaper article, Point Reyes, California.

San Francisco Examiner. 1984. Gill nets kill hundreds of birds. 1 August, 1983. Newspaper article, San Francisco, California.

Takekawa, J. E., H. R. Carter, and T. E. Harvey. 1988. Decline of the common murre in central California. San Francisco National Wildlife Refuge. Unpublished manuscript.

This World. 1986. Gill nets. 14 September, 1986. Newspaper article, San Francisco, California.

Questions and Concerns

READING 2

As you read this article consider the following:

This is the fisherman's perspective of this issue. Do they have genuine concerns? Are they united against this issue?

California passes gillnet ban

By Mick Kronman
Field Editor

Southern California's century-old inshore gillnet fishery died Nov. 6, a victim of a state-wide initiative.

By a margin of 55.5% to 44.5%, California voters passed Proposition 132, amending the state's constitution to include the so-called Marine Resources Protection Act of 1990. Its provisions feature:

- A ban on gillnets and trammel nets within 3 miles of the California coast from Point Arguello to the Mexican border (a 250-mile stretch), and for 1 mile or 70 fathoms, whichever is less, around the region's eight channel islands. (Further, it will add to and make permanent a broad mosaic of existing gillnet restrictions and closures in central and northern California.)

As a result, the commercial harvest of halibut, angel shark, white seabass, kingfish, barracuda and yellowtail will be reduced or eliminated. Inshore gillnetters in affected areas must quit the business by Jan. 1, 1994.

- A ban on rockfish gillnets in all areas, even outside the 3-mile and 1-mile zones. (This provision is a preemption of federal jurisdiction, and it's unclear whether it will pass legal tests.)
- Compensation (after enabling legislation is passed) to gillnet fishermen who surrender their permits. Estimated to cost $3.4 million overall, the bailout will be funded by a fee hike for gillnet permits during the next three years and a $3 surcharge on sport-fishing licenses. Compensation will equal a fisherman's average yearly ex-vessel landings in affected areas from 1983-1987. No compensation is offered for gear or boat devaluation.
- A requirement that the California Department of Fish and Game (CDFG) establish four new "open water" ecological reserves for marine research. Sport fishing may or may not be allowed in the reserves, depending on future state action.

Assemblywoman Doris Allen is architect of Proposition 132. She chairs the Committee to Ban Gillnets and is a long-time adversary of California's commercial fishing industry. She embraces the election results, which will toss about half of the state's 450-plus licensed gillnetters out of work or into other fisheries.

"I'm euphoric," Allen says. "Voters read the ballot arguments and came to a logical conclusion. They see with their own eyes that gillnets are indiscriminate killers. They see sea lions with gouges in their necks. They watch our marine resources being depleted. They believe this kind of devastation isn't worth the price. They believe this is good government."

Allen says the measure earned a spot on the ballot because almost 1 million petition signers wanted it there. The measure belonged therc, she adds, "because commercial fishermen's powerful lobby in Sacramento" had twice thwarted parallel legislation.

While the assemblywoman and her supporters (primarily yacht brokers, sport-fishing groups and tackle manufacturers) celebrated, commercial fishermen mourned. At some ports, men huddled in small groups, mumbling fears that sport fishermen are spreading an anti-commercial virus that could infect the whole fishing industry.

"You can bet Doris Allen is not finished. In fact, she's just getting started," predicts Mike McCorkle, a California Gillnet Association (CGA) director and former *National Fisherman* Highliner. "I think her ultimate goal is to eliminate inshore commercial fishing from Point Arguello to the Mexican border."

Allen says she's "not anti-commercial, but pro-environment." McCorkle, for one, doesn't buy that explanation.

"Doris and her supporters want pleasure fishing only; it's as simple as that. And don't think the problem is just here. Commercial fishermen on all coasts are under fire. No region and no fishery is safe. Its hard to say who will be next — could be draggers, purse seiners, trappers or longliners."

"It could happen anywhere, especially since sport fishermen are now probably drunk with power," adds Rob Ross, executive director of the California Fisheries and Seafood Institute (CFSI), an industry lobbying group. "These people won't be happy until there's a ribbon of water surrounding the United States where commercial fishing is prohibited."

Allen admits her agenda is yet unfulfilled. "We need to fight against any type of commercial fishing gear that is as environmentally damaging as gillnets, such as trawl nets," she says. "You can bet gillnet fishermen and their allies will seek to get back into our coastal waters with equally efficient killing machines."

What Happened

For fishermen, Proposition 132 capped a years-long anti-gillnet movement they say will cost 1,000 offshore jobs and 3,000 onshore jobs. "We'd already been hammered senseless by rhetoric, emotionalism and lies in the press and in public; this was just the last coffin nail," one frustrated gillnetter explains.

"I was exceptionally disappointed at the lack of commitment among fishermen to fighting this thing. It was a winable battle, but fishermen didn't do the work that had to be done — handing out leaflets, educating the public."

"That's a problem with the initiative process," he adds. "Twice, legislators saw through [Allen's] bull. They threw out her anti-gillnet bills, saying the proposals lacked supporting biological data. But when she went public with her cause, there was no requirement that she had to be honest. And she wasn't."

Industry representatives say sensationalism, not science, fueled the anti-gillnet campaign. Proposition 132 literature says gillnets ". . . stretch for miles, mutilating and killing any marine animal that becomes entrapped." Fishermen disagree, insisting inshore nets are highly discriminating since they're regulated by mesh size and length.

In California ballot arguments, Allen and her supporters say, ". . . 72% of all fish species caught in gillnets have absolutely no commercial or economic value. They are caught and killed in nets, then simply thrown back into the sea to rot . . ."

Fishermen, however, cite CDFG studies in the mid-1980s that indicate the opposite — 83% of all species caught in inshore gillnets are utilized or released alive. Markets for bycatch such as skates and rays (for crab bait) and crab claws are on the rise, and little of what's taken in nets is ever wasted, they note.

When gathering signatures during the initiative drive and stumping before the election, Proposition 132 supporters placed posters in dozens of communities that read, "Save the Whales, Save the Dolphins, Stop the Slaughter."

Fishermen say this was a deliberate attempt to sell the initiative on fraudulent grounds, confusing inshore gillnet issues with non related matters such as high-seas driftnets and tuna/porpoise interactions.

Anti-gillnetters said thousands of harbor seals and sea lions die annually in near-shore nets, and they aired film and photos of sea mammals suffering from monofilament mesh wrapped around their necks.

Fishermen and mammalogists say the sea lion deaths are caused by skyrocketing population of the pinnipeds (about 90,000 in California, growing at 5% per year), not a perceived increase in gillnetting. They point to sea lion "invasions" in San Francisco Bay and Puget Sound as examples of how far and wide the animals have roamed.

Net-trapped mammals usually get in trouble only when raiding the gear, adds one Santa Barbara gillnetter. "In some areas, they eat nearly 100% of a day's halibut or seabass catch. They're getting a much bigger share of the fish than us."

On the whale issue, fishermen note National Marine Fisheries Service (NMFS) statistics indicating that fewer than five gray whales (out of 22,000 migrating along the California coast) die annually in gillnets. Furthermore, they say commercials, not sports, crafted whale-protection legislation three years ago, requiring heavy anchors and breakaway panels on the nets.

Lawmakers, fishermen say, have thus adequately managed the gear. They say Proposition 132, on the other hand, was an end-run around normal legislative procedure and a misuse of the voter initiative process.

Initiative proponents said commercial fishermen take more than a fair share of the resource. Commercials, however, note that sport landings of halibut are near all-time highs, and yellowtail, bonito and barracuda angling has been above average in recent years.

"To say Doris Allen and her supporters misrepresented the facts and snookered the public would be an understatement," says CFSI's Ross. "To say she's an environmentalist is a joke. She used fabricated resource issues to hide the truth — that sports want the fish for themselves. "She also got lucky. We only lost by 10% [the vote was closer in inland counties]. Since they outspent us by more than 10-1, I'd say they benefited from the fact that high-seas driftnets, tuna nets and wayward whales were in the news a lot during the past year."

Not Enough Involvement

Some gillnetters and processors active in the anti-initiative campaign partially blame fishermen for the outcome. Too few, they say, were willing to spend time and pay assessments on their catch to battle the measure.

"Fisheries groups nationwide need to rally against issues like these. It doesn't matter whether you fish pompano in Florida, tuna in the Atlantic or halibut in California. We're in the same predicament, and we need to hang together."

"I was exceptionally disappointed at the lack of commitment among fishermen to fighting this thing," says Craig Ghio, owner of Ghio Seafoods in San Diego. "It was a winable battle, but fishermen didn't do the work that had to be done — handing out leaflets, educating the public.

"We lost because fishermen didn't spend time on the campaign, not because they didn't spend money.

"Fisheries groups nationwide need to coordinate their efforts and rally against issues like these. It doesn't matter whether you fish pompano in Florida, tuna in the Atlantic or halibut in California. We're in the same predicament, and we need to hang together."

Ross says he hasn't ruled out the option of fighting Proposition 132 in the courts. "I'll have to wait and see what industry wants to do, but I feel there are some serious legal issues here, including equal protection under the law."

As it stands, Proposition 132 will give 4% of California's population — recreational-fishing license holders — 100% access to the state's marine resources. It is also a preemption of federal jurisdiction since rockfish grounds are of ten in federal waters."

Zeke Grader, head of the Pacific Coast Federation of Fishermen's Associations, isn't sure court action is the answer. "We may be better off taking our lumps, cleaning up our fisheries and structuring legislation to address any problems, " he says.

"We can also call to task all the people and organizations who claim to protect the marine environment but who simply seek to eliminate competition," he says. "For example, are fish derbies, where people just play with the resource, an intelligent use of the marine environment? What about treble hooks that stick in the bellies of short halibut? Or mooching for salmon with small hooks? What about environmental review? Has damage to sport-fish stocks by continual pressure been adequately investigated?"

On the up side, says Grader, several potent environmental groups, including the Sierra Club, GreenPeace, Ocean Alliance and Friends of the Sea Otter, did not support Proposition 132. "This could pave the way for negotiations with these groups in the future," he notes.

But in the here and now, many fishermen appear stunned and rudderless. "The election left me completely discouraged about my life," says Hank Hubble, a Ventura gillnetter. "Where can we go? Sea urchin, abalone and salmon fisheries are closed; most of our boats are too small (25' to 40') to drag; purse seining requires huge capital investment; and trap fisheries are already crowded. On top of that, no alternatives to gillnets work as cleanly or efficiently for taking inshore fish. None."

On the marketing side, processors are gearing for the worst. "I'll have to reorganize my entire business," laments Mike Wagner, owner of Seafood Specialities in Ventura. "Eighty percent of the fish I buy —more than $2 million worth per year — comes from inshore gillnets. That's a $6 million loss to the local community."

Wagner says he can buy halibut and seabass from Mexico and rockfish fillets from Canada, but he'll receive less fish overall, and consumers will pay 50% to 100% more at the seafood counter. Futhermore, quality could slip. "Instead of buying from day-boats, I'll have to buy three-day-old Mexican fish," he explains.

Wagner says some fresh fish supplies, such as 300,000 lbs. of angel shark per year, will drop from the market altogether. "And crab claws, a marketable gillnet bycatch that's sold in markets throughout Los Angeles, will also disappear." □

Discussion for Gillnet Fishing on the West Coast

1. What is the 'dead bird' project?

2. How effective is the gillnet and describe how it is employed.

3. Explain how the Federal Law was being broken by gillnetters.

4. Do you think the gillnet ban will have a beneficial impact on marine life? If so how much of an impact do you think it will be?

5. Review the list of supporters of the gillnet ban. Do you see a trend? Who is in support of this law and who do you not see listed? (I wonder if that oil company that saves little foxes on TV donated to this cause?)

6. Why do you think there was some dissension among the ranks of fishermen?

7. If you represented the fishermen what position would you take on this issue?

Introduction

Rivers, Rivers, Rivers, and Not a Drop to Drink

This exercise looks at the problems facing rivers of the Northwest. The conflicts between those who want to use this resource are a major problem pitting the economy against the environment and the need for power against the fisherman.

San Francisco Bay has had 62% of its fresh water diverted to farming and to the parched southern half of the state. As a result of this diversion a 60-80% decline in striped bass production has been noted. California has had a 90% reduction in wetlands and a 65% reduction in salmon numbers. The Columbia River Basin has suffered a 40% loss in spawning areas and as a consequence an 80% loss in steelhead and salmon has been recorded. West Coast salmon are about to be declared endangered under the Endangered Species Act and in Idaho the Shoshone Indians want the Snake River Sockeye so listed (National Fisherman, 1991 Yearbook). So what is happening?

Drought plus water diversion for farming and drinking have had a serious impact upon water levels and water quality. Southern California needed drinking water and even with rationing in some parts of the state it needed additional supplies from up north. As a result the Klammath and San Joaquin Rivers were tapped to their maximum. What was left was heavily polluted with farm run off, sewage and industrial waters. Selenium levels increased in the remaining water and at the Kesterson National Wildlife Refuge in California, a mass die-off of migratory wildfowl occurred. Emergency measures to drive away incoming wildfowl occurred and most of the tainted pools were drained and filled. The only solution to this problem seemed to be the withdrawal of agriculture lands from production, the ceasing of the application of fertilizers and pesticides and the continual flushing of these marshes and rivers with fresh, clean water. Pollution occurs from many sources. Some of the major culprits are livestock manure, fertilizers, pesticides, mining wastes, lawn chemicals, construction erosion, and city dwellers waste oils and auto fumes. Amendments to the Clean Water Act (1987) address the problem of contaminated runoff pollution by asking states to voluntary programs. However the $400 million authorized for the programs has not been released.

The El Niño condition off the coast of the western U.S. has altered the ecology along the coast. The normal small fish and zooplankton fauna have undergone a change and with the abnormally warm water it seems that the salmon may have been adversely affected. The Point Reyes Bird Observatory administration has charged that the mass of boat traffic, engine noise, diving etc. has had a definite impact upon the birds in the area. Perhaps these factors have also affected the fish populations?

Steelhead runs are at their all time low and at a recent symposium in Washington state many other reasons were discussed. Mid ocean driftnets may have intercepted the runs but no data is available. The El Niño condition was a good prospect and so was the potential for competition between wild and hatchery stocks and perhaps diseases had been introduced by released hatchery or captive bred fish? Marine mammals have been blamed particularly in light of the the problems at the Ballard Ship Locks in Puget Sound, Washington. Here at the Locks, many dozens of sea lions have learned to catch the migrating Steelhead navigating the restricted passage. What can be done in this situation? Who has precedence, the fully protected sea lion or the rapidly disappearing salmon? But the main culprit was believed to be the fact that the rivers have been tamed. Dams have stopped the uncontrolled flow of storm water and in so doing benefitted the community with an abundance of electric power and the fish have suffered. Fifty-nine years ago the Columbia River had its first dam built across it. Sixteen million salmon were to be found in its free flowing water. Today the estimate is two and one half million fish. Spawning grounds have been buried and other areas drained. Oregon, Washington and Idaho pump out hundreds of millions of acre feet of water to serve agricultural needs and the water that is left is warmer and more polluted.

Pollution comes in many forms but the major headaches in Oregon seem to be coming from radioactive leachate from the Hanford nuclear site, mining and logging activity and of course agriculture runoff. At the moment state and federal regulators seem incapable of keeping up with the mass of illegal activity. As you read the following articles consider how you would address this problem if you had the power to do so.

Barry Thomas

Questions and Concerns

READING 1

When reading this article consider the following:

What is being planned to save the fish? Does it seem practical?

Reading 1

Death of a river (No. 3 in a series)

Bill Jennings and Stan Thomas

The once prolific fishery of the Mokelumne river has been nearly destroyed. In times past the Mokelumne river supported a fall spawning population of 60 to 80 thousand Chinook salmon and Steelhead trout. But last fall only 400 salmon were counted at Woodbridge Dam in Lodi.

Two factors account for the loss of the fishery: the depleted waterflow and the enhanced level of toxics in the river. The problems stem from the East Bay Municipal Utility District's (EBMUD) negligent stewardship of the Mokelumne River.

The previous installment discussed the issue of water quality, specifically EBMUD's continued operation of the poisonous Penn Mine Waste Treatment Facility. This installment discusses the critical issue of EBMUD and water flow in the Mokelumne river. At the outset it is important to realize that the struggle is not between the needs of people versus fish. There is water for both. It is a struggle between selfish institutional arrogance and a decent respect for the public trust and the rights of generations yet unborn.

Since salmon have short life spans and spawn once, they need conditions favorable to successful spawning, egg incubation, rearing and migration in order to ensure the survival and genetic diversity of the species. Adequate water flows at appropriate times is absolutely crucial to this natural process.

Before EBMUD began filling Camanche Reservoir in 1963, they reached a highly political agreement with the Department of Fish and Game (DFG) on measures to "protect" downstream fisheries. A pitifully inadequate one and one-half percent of the Mokelumne's annual average 730,000 acre-foot runoff was reserved for the downstream aquatic ecosystem. The result is a fishery that barely survives.

Hope for the Mokelumne River remains alive, however. State legislation passed in 1988 mandated a goal to significantly increase the natural production of salmon and steelhead trout by the end of this century. DFG's initial step to develop solutions to fish and wildlife problems in the Mokelumne River was a report released to the public in late February, "Lower Mokelumne River Fisheries Management Plan."

DFG's management goals for the lower Mokelumne River are to optimize chinook salmon, steelhead trout, American shad, and striped bass habitat conditions and populations. Chinook salmon are emphasized due to their significant value to sport and commercial fishing interests.

To provide adequate water flows, the DFG management plan calls for approximately 28% of the average annual unimpaired flow in the Mokelumne to be committed to fisheries. Specific temperature requirements to protect fisheries are included. Stringent water quality standards for dissolved oxygen, pH, toxicity, and turbidity are recommended.

The condition of the Mokelumne River is symptomatic of a much larger problem: gross misallocation of California's scarce water resources. Agriculture uses 85% of the state's water; urban residents, industry, wildlife, and aquatic life share the remaining 15%.

Agricultural water is heavily subsidized by the Federal government. A typical urban dweller pays $300 per acre-foot of water; a typical agricultural user pays $2.50 for the same amount of water. Consequently, water waste horror stories abound of California farmers using heavily subsidized water to grow surplus crops in a semi-arid climate. Irrigated pasture uses more water in a normal year than all urban citizens combined. In California's $600 billion economy, one-ninth of all the water in the state is being used to support an industry that provides 1/6000 of the economy. Also, in non-drought years, California's rice crop alone diverts about 2.5 million acre-feet of water each year - enough for 25 million urban users or, conversely, for dramatically enhanced fish and waterfowl habitat.

Forty percent of all the water in California is controlled by the Federal Bureau of Reclamation. EBMUD's policies on water use are bad. Those of the Federal government are far worse. Write your Representative anyone?

Connections readers may obtain a copy of the booklet, *East Bay MUD, Penn Mine, and the Death of a River* by sending $1.00 with your request to Committee to Save the Mokelumne, 2800 W. March Lane, Suite 326, Stockton, CA 95207.

Courtesy of *Connections*, published by the Peace & Justice Network of San Joaquin County, P.O. Box 4132, Stockton, CA 95204.

Reading 2

Questions and Concerns

READING 2

When reading this article consider the following:

How is the pollution getting into the aquifer? Where did it come from?

Radioactive water inches toward Columbia

Hanford engineers figured the uranium would stick to the soil as the water went down. Then somebody made a mistake...

By JAMES LONG
of The Oregonian staff

Like most halfway decent ideas at the Hanford nuclear site, this one lasted about as long as it took somebody to make a mistake.

Opened the wrong valve, the story goes.

And now, a generation later, officials are still debating what to do about it.

The outcome of this human lapse was a 5-square-mile underground plume, approximately, of radioactive and chemically contaminated water.

The plume is inching its way slowly toward the Columbia River, destined to reach it by the year 2185, according to computer studies. Unfortunately, the plume, spiked with radioactive uranium-238, technitium-99, iodine-129 and some other stuff, will not be any less dangerous when it does reach the river.

The Oregon Hanford Waste Board, a 12-member state panel that guards Oregon's stake in the Hanford cleanup, is worried about the plume.

The board sent a letter to federal and Washington state officials in charge of the cleanup, asking them to put it on an "expedited response" list. This put the plume high on the list of things to do in what is likely to be a multigenerational, multibillion-dollar scrubbing of the former military atomic site, which occupies more than 500 square miles of sagebrush just outside Richland, Wash.

The board is concerned, ironically, that some of the cleanup processes now under way at Hanford could have the unintended consequence of forcing the radioactive plume faster toward the Columbia. The river runs for hundreds of miles along Oregon's northern rim.

Not everyone agrees that the waste board's fear is well-founded, however.

Dr. Dick Belsey, an emeritus professor of pathology at Oregon Health Sciences University and a member of the waste board, sounded the alarm on the plume. He's afraid that new discharges of water into the ground near the plume, in connection with another urgent cleanup project, may shove the existing contaminants faster toward the river or at least spread them out and make them harder to retrieve.

The focus of Belsey's concern is a flow of water into a gigantic in-ground drain field known as a "crib," about 15 miles from the Columbia River. The crib is near a uranium processing complex that will be used temporarily to convert hundreds of thousands of gallons of uranium-laden nitric acid into a relatively safe powdered uranium trioxide.

During World War II, the government built three reactors at Hanford to make plutonium for bombs. Plutonium is produced in tiny amounts by stacking uranium-238 in a reactor and allowing it to bombard itself with its own flying neutrons. Some of the neutrons get tangled up in the uranium atoms, which end up as the slightly heavier element, plutonium.

Because the military was interested in plutonium, rather than uranium, much of the leftover uranium was dissolved in acid and pumped into waste tanks. There it remained until the 1950s, when the government began running short of mined uranium.

U-plant, one of Hanford's original reactors, which had never operated, was taken out of mothballs to recover uranium from the waste tanks.

The recycled uranium concentrate was piped into a nearby plant, known as UO3 Plant, and converted to uranium trioxide, an orange-colored powder that could be made into new reactor targets to make plutonium.

But the scientists had a slight problem:

When they took the uranium out of the stew from the tank farm, they also ended up with a lot of junk they didn't want, including lots of uranium-contaminated water. They couldn't put all the water back into the tanks because there wasn't room.

The solution was to build a couple of huge cribs, U-1 and U-2, and put it in there. In the ground.

It wasn't quite as crazy as it sounds: Hanford's engineers figured that the uranium would stick in the upper layer of soil, which would filter the water as it went down.

And that's just about what happened until somebody made a mistake one day.

Instead of sending water into the cribs, somebody accidentally unleashed a torrent of acid cleaner. The acid stripped the uranium out of the soil and sent it plunging toward ground water.

As soon as the mistake was discovered, of course, the acid was shut off and nothing else was put into U-1 or U-2.

The engineers built a new crib, U-16. And only clean water was discharged into U-16.

Then something else went wrong.

As the water percolated down from U-16, it hit a clay layer in the soil that acted like the bottom of a pan, causing the water to spread sideways. As it spread, it picked up the now-mobile uranium from U-1 and U-2 and began moving it again. It also carried radioactive technicium and iodine and other chemicals.

The spreading plume then found a well-hole in the area — a well the someone had drilled for an entirely different purpose — and the hole provided an expressway to ground water.

Hanford officials discovered this new mishap, to their horror, and tried in the summer of 1985 to pump the plume out of the ground and clean it.

It was quite a job.

U-1 and U-2 had received more than 4 tons of uranium, much of which was now swimming around in the groundwater, as the scientists learned to their consternation when they measured the plume's radioactivity.

The ground water near the cribs showed a uranium content producing 80,000 picocuries per liter, for instance. The proposed U.S. drinking-water standard is 30 picocuries.

After a summer of pumping and cleaning, Hanford crews got the uranium readings down to 20,000 picocuries. Then the results leveled off, and pumping was suspended as winter bit the sagebrush. Next spring, it was decided that not much more could be done, so the pumping was abandoned.

The conventional wisdom at Hanford is that the pumping project was at least a qualified success. But Belsey, of the Oregon board, wrote a note to other members saying he was concerned that "significant parts of the plume may no longer be in the area being monitored. Also, contaminants may have moved down in the aquifer, out of reach of existing monitoring wells."

A new crib, U-17, replaced the U-16 fiasco in 1987 and has received nearly 500,000 gallons of clean wastewater, without known mishap, from Hanford cleanup operations.

U-17 mostly gets steam distillates from a temporary program to get rid of the most dangerous materials around the uranium processing complex.

The laws and court orders that govern Hanford's ground-water discharges allow U-17 to receive up to 10 clean gallons of water a minute for this purpose.

But Belsey and the Oregon Hanford Waste Board are leery about the gallon limit. They point out that it's an averaged figure and that it would be possible to stay within the average and still release huge slugs of water from time to time, possibly affecting the plume.

But Hanford officials don't think Belsey's concern is well-founded. They also point out that some further discharges into U-17 are necessary to relieve much worse hazards around the uranium complex.

Bill Ross, who manages the UO3 plant for Westinghouse Hanford Corp., says the cleanup of about 200,000 gallons of uranium-bearing nitric acid during the next three months will depend on U-17.

The dangerous acid-uranium brew is stored in two stainless steel tanks at the plutonium and uranium extraction plant — called the Purex plant. The tanks are long past their design-life and rest on tired concrete foundations that show ominous cracks.

Ross says Purex will separate the liquid into 200 metric tons of uranium concentrate and 50,000 gallons of nitric acid and also will boil off about 150,000 gallons of water.

The uranium concentrate will go to UO3 for conversion to uranium trioxide powder; the acid will be set aside for re-use in cleaning radioactive residue out of old processing machinery; and the boiled-off water, mildly acidic but otherwise clean, will be pumped in batches into a 12,000-gallon holding tank and checked for contaminants.

From there, it will be sent in smaller batches to a 500-gallon feed tank, where it will be neutralized and re-checked before being pumped into the U-17 crib at the rate of 100 gallons per minute. The small feed tank is pumped about every 12 minutes until the larger tank is emptied, a process requiring maybe six hours.

Once the uranium-acid liquids are taken care of, the involved plants themselves will be cleaned up and shut down, probably forever.

And in 1995, the post-cleanup trickle of about 2 gallons per minute that will be allowed for U-17 will be terminated, and all future water discharges will go through a new treatment plant.

In the meantime, the never-ending controversy over how best to clean up Hanford, and how not to, frazzles even the most battle-tested officials.

John Hunter, the U.S. Department of Energy's assistant manager for waste management at Hanford, says the question of U-17's impact on the uranium-water plume was studied and discounted by technical experts, mainly on the grounds that U-17 is downhill from the plume.

Water, he says, does not flow uphill, even at Hanford.

"We're certainly concerned about the water discharges, too," Hunter groused.

"We have reduced water discharges to the ground on this site by over 80 percent, and that trend is continuing. We are operating, we believe, in strict conformance with the law, in order to do something intelligent. And that is to get this facility closed down and put it out of its misery, OK? But every time I turn around, I have somebody wanting to prolong the agony, because they have some strange notion that their agenda may be more overriding than the agenda we're on."

Then Hunter laughed. "I guess you can see I get irritated."

Discussion for Rivers, Rivers, Rivers, and Not a Drop to Drink

1. Should the Endangered Species Act impact our living standards?

2. Energy rates in the Northwest are probably the lowest in the nation thanks to hydro-electric power. Is it time to pay the piper and raise prices to save wildlife?

3. Consider the pre-construction of the Camanche Reservoir agreement. Do you think it gave adequate consideration for the fish and wildlife?

4. Can we afford to subsidize the farmer?

5. If you had to prioritize the enforcement process, which source would be your first... which last?

Mining Update

Update on Raven/Berg/Johnson *Environment*, p.187

Updated information on some environmental aspects of coal mining, as discussed on page 187 of *Environment*, was not able to be placed in the first printing. The authors would like to direct the reader's attention to the following changes:

column 1: The 90,000 mining fatalities recorded in the 20th century should be viewed in a historical perspective. There have been great improvements in reduced fatalities due to the efforts of mining companies and to federal regulators in enforcing safety standards. According to the U.S. Department of Labor, 54 coal miners died during 1992, which is the second lowest total in history. Annual accidental deaths of coal miners numbered from 2000 to 3000 in the earlier part of the century.

column 1: There are more than 2000 deaths per year in the United States resulting from black lung disease (a form of pneumoconiosis), although it is difficult to determine precise numbers. Improved ventilation equipment and other safety measures may result in a lower incidence rate of this disease in the future; the matter is under study by coal companies and government agencies.

column 2: As discussed on page 330 of *Environment*, the 1977 Surface Mining Control and Reclamation Act (SMCRA), requires coal companies to restore areas that have been surface-mined, in order to prevent the past practice of large open pits left after surface mining. SMCRA also requires that topsoil be replaced after mining, and that procedures to control erosion and comply with water quality criteria are established.

Note: Look for more coverage of reclamation of surface-mined lands in the second edition of *Environment*.

Professor and Student Comments

Your Feedback Counts!

The *REGIONAL ENVIRONMENTAL ISSUES MANUALS* are a "living" project which will be growing and changing with editions, like our environment. We at Saunders College Publishing are delighted to be pioneering such new and unique supplements for our environment textbooks, as a way of bringing regional issues closer to your home. Your feedback counts! Please detach, complete and return this form so we can continue to update our manuals and cover the environmental issues that are important in your region.

Your Name ______________________ Are you a Student ______ or Professor ______?

School Affiliation and State ______________________

Address ______________________

City ______________________ State ____________ Zip ____________

Phone Number Home: ____________ Office: ____________

Which Regional Environmental Issues Manual do you use? (Circle one)

Northeast	Southeast	Mid Atlantic	Great Lakes
Midwest	Southwest	Northwest	Canada

What issues in this manual were most interesting and useful to you in class? ______________________

____

What issues can be improved? ______________________

____

Are there issues/topics which are not included which should be included in the next edition? ______________________

____

____

Are there issues/topics which should be deleted from the manual? ______________________

____

What additional sources (articles, readings) would you recommend for inclusion in the next edition? ______________________

____

____

Do you feel your state is placed in an appropriate region, one in which other states of the region have similar environmental problems and concerns?

Yes ______ No ______ If not, what region should your state be in? ______________________

How would you rate the Regional Environmental Issues Manual as a free supplement for your text:

____ Excellent and very informative

____ A good regional supplement to our text

____ Average

____ Needs to be improved in the future

____ Poor

____ Other (Please explain) ______________________

For Instructors:

Would you be interested in working on a revision of this or other Regional Environmental Issues Manuals? Yes ______ No ______

Other Comments: ______________________

____

____

Please complete the other side of this form, detach and return.

FOLD HERE AND TAPE CLOSED

PLEASE
PLACE
POSTAGE
HERE

Please circle the manual you use:
Northeast
Southeast
Mid Atlantic
Great Lakes
Midwest
Southwest
Northwest
Canada

Return to:

Regional Environmental Issues Manuals
Julie Levin Alexander, Senior Editor
Saunders College Publishing
The Public Ledger Building
620 Chestnut Street, Suite 560
Philadelphia, PA 19106